SPCC and Business of Cleans 2021

Online
8-10 June 2021

ISBN: 978-1-7138-3096-2

Printed from e-media with permission by:

Curran Associates, Inc.
57 Morehouse Lane
Red Hook, NY 12571

Some format issues inherent in the e-media version may also appear in this print version.

Surface Preparation and Cleaning Conference (SPCC)
c/o Mark Thirsk
Linx Consulting-PO Box 384
Mendon, MA 01756-0384
USA

Phone: 617-273-8837

mthirsk@linx-consulting.com

Additional copies of this publication are available from:

Curran Associates, Inc.
57 Morehouse Lane
Red Hook, NY 12571 USA
Phone: 845-758-0400
Fax: 845-758-2633
Email: curran@proceedings.com
Web: www.proceedings.com

TABLE OF CONTENTS

Hilltop Economics

What The Global Economic Outlook* Implies for Semiconductors

June 8, 2021

Duncan Meldrum
Chief Economist, Hilltop Economics, LLC

duncan.meldrum@hilltopeconomics.com

Not to be distributed without prior permission

**Global and U.S. outlooks based on Consensus Economics, Inc. CONSENSUS FORECASTS®*

Hilltop Economics

Summary Overview

- **Economic Recovery Underway**
  - **Fueled by stimulus packages**
  - **Inflation the primary risk**
- **Consensus Outlook Continues to Rise**
  - **Continued government support**
  - **Strong consumer demand**
  - **Strong business demand**
- **Upstream Semiconductor Markets Remain Strong**
  - **Anticipate some softening once the pandemic recession fades.**

Hilltop Economics

Growing Optimism in Consensus Economic Outlook

- Pandemic is getting controlled in key economies
  - Confidence rises
  - Jobs recover
- Stimulus leads to investment & productivity gains
  - Inflation (sustained price increases) does not develop
  - Current price spikes just temporary adjustments
  - Supply chain issues subside
- Institutional backdrop stable (lower uncertainty about economic policy)

SPCC Presentation
June 8, 2021

3

Hilltop Economics

World* Real GDP Growth
2019: 2.4% 2020: -3.8% 2021: 5.6% 2022: 4.3%

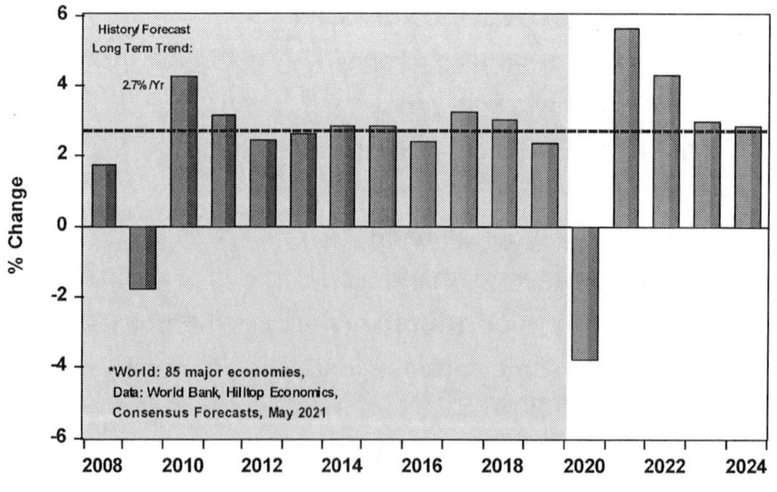

SPCC Presentation
June 8, 2021

4

Hilltop Economics

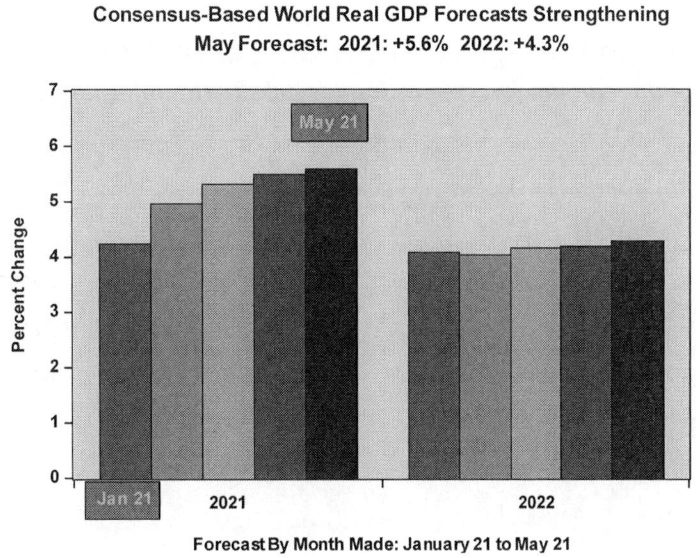

SPCC Presentation
June 8, 2021

5

Hilltop Economics

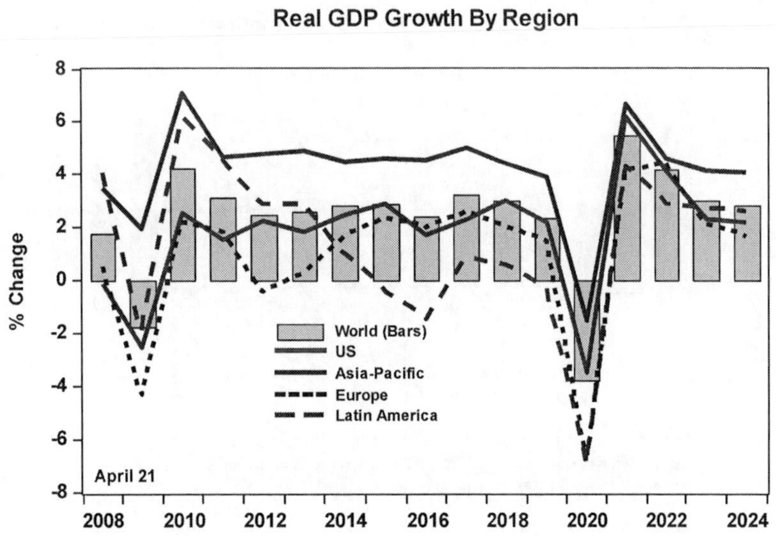

SPCC Presentation
June 8, 2021

6

Hilltop Economics

Fiscal (Govt Spending) Stimulus Guarantees Recovery in 2021

Helps explain regional forecast as it adds risks (debt, inflation) to the future

Govt Debt % GDP	2019	2020	*2020 % Increase*	2021	*2021 % Increase*
Total World	*83.7*	*97.3*	**+16.2%**	*98.9*	**+1.6%**
USA	*108.2*	*127.1*	**+17.5%**	*132.8*	**+4.5%**
China	*57.1*	*66.8*	**+17.0%**	*69.6*	**+4.2%**
EURO Area	*84.0*	*96.9*	**+15.4%**	*98.0*	**+1.1%**
Japan	*234.9*	*256.2*	**+9.1%**	*256.5*	**+0.1%**
Latin America	*64.8*	*77.7*	**+19.9%**	*75.9*	**-2.3%**
Src: IMF Fiscal Monitor April 21					

SPCC Presentation
June 8, 2021

Hilltop Economics

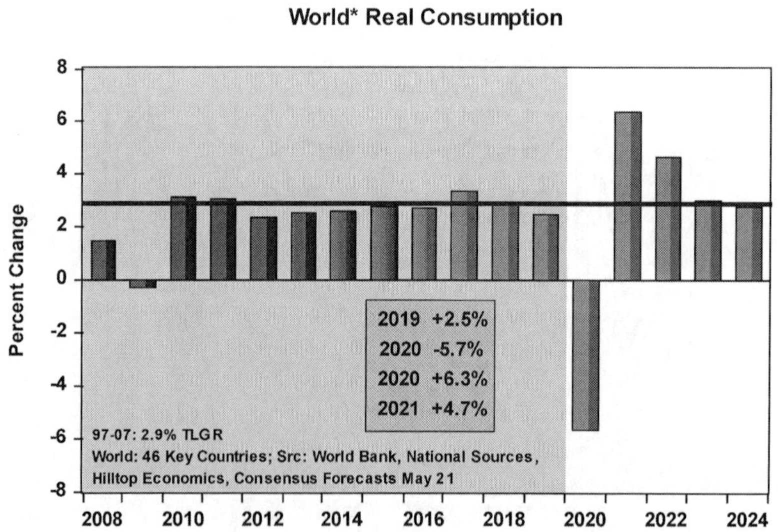

SPCC Presentation
June 8, 2021

Hilltop Economics

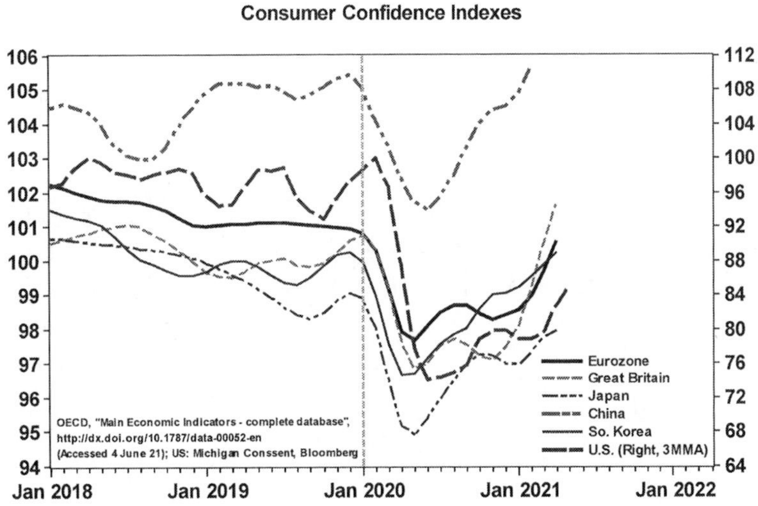

SPCC Presentation
June 8, 2021

9

Hilltop Economics

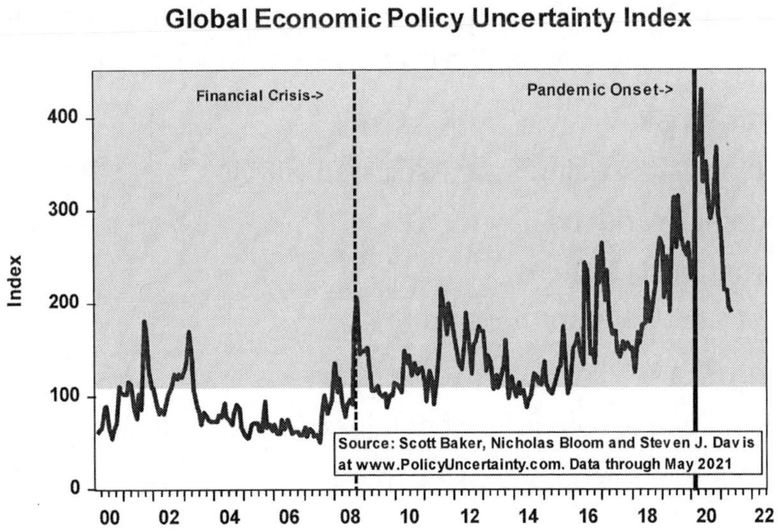

SPCC Presentation
June 8, 2021

10

Hilltop Economics

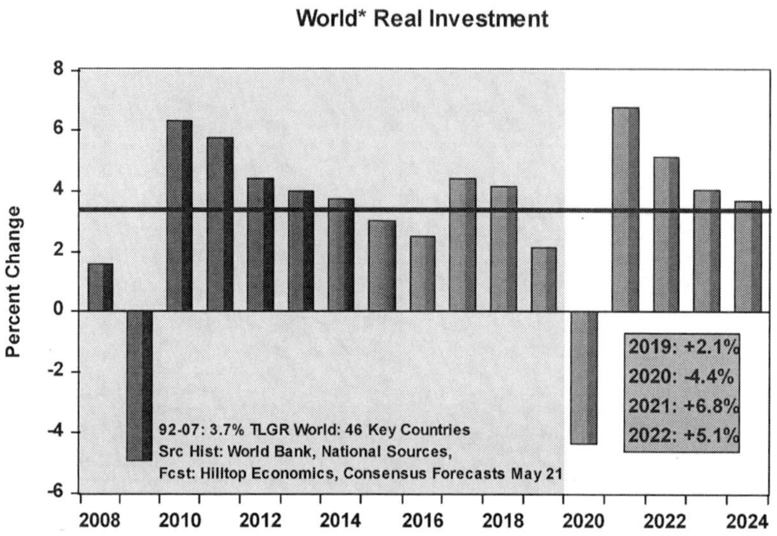

SPCC Presentation
June 8, 2021

11

Hilltop Economics

What Could Disrupt the Consensus Economic Outlook

The main risks that could cause the forecast to be wrong

- Downside risks
  - Inflation & cascading risks associated with it
  - Pandemic (variants)
  - International relations
  - Investment does not develop
- Upside risks
  - Post pandemic technology boom boosts productivity & economic growth

SPCC Presentation
June 8, 2021

Hilltop Economics

The Inflation Risk Story: U.S. Focus

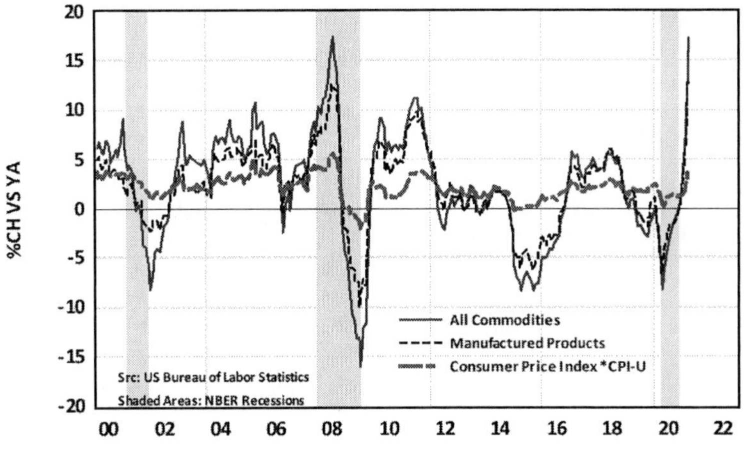

Why?

- Commodity prices are sharply higher
- Product prices rapidly follow,
- Supply chain disruptions causing tightness
- Consumer prices jumped sharply (4.2% year-over-year in April) so flow-through more than in past
- Federal Reserve policy will keep money "easy"
- Fiscal stimulus keeps demand strong
- Strong final demand with tight supply will lead to sustained inflation

SPCC Presentation
June 8, 2021

13

Hilltop Economics

Inflation Risk: Consensus Doesn't Think So...

CPI Inflation	2020	2021	2022
US - CPI	**1.2**	**2.8**	**2.3**
– PPI	**(1.3)**	**5.3**	**2.2**
Eurozone	0.3	1.7	1.3
UK	0.8	1.6	2.2
Japan	0.0	0.0	0.5
China	2.5	1.5	2.2
So Korea	0.5	1.7	1.5
Taiwan	(0.2)	1.5	1.4
Brazil	4.5	4.9	3.8
India	6.2	5.0	4.8

Why Not?

- Year-ago-comparisons are to distorted pandemic shut-down months (April-May-June)
- Supply chain disruptions are temporary: supply-demand imbalances will ease after pandemic passes
- Capacity utilization remains low
- Fiscal stimulus spending is temporary
- Employment rates are still low
  - US: employed as % of adult population still BELOW worst of financial crisis recession
- EXPECTATIONS remain low
  - Behavior has not changed
  - No broad panic spending
  - Cost escalation indexes not endemic
  - Wage increases not excessive relative to productivity
- Central banks know how to fight inflation – and will step in "in time"

SPCC Presentation
June 8, 2021

14

Hilltop Economics

Inflation Risk Seen As Low By Key Market Indicators

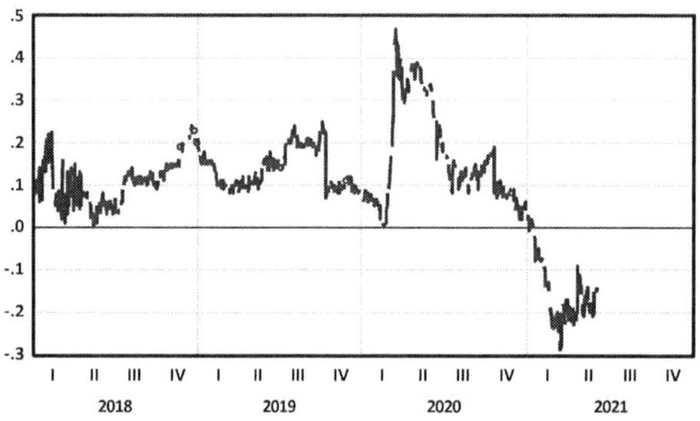

10-Year TIPS-based inflation forecast MINUS 5-Year TIPS-based forecast.

A negative indicates the market's most informed and serious agents expect the inflation spike to be transitory.

Current 10YR: +2.40%
Current 5YR: +2.54%

Caveat: Most informed and serious agents have never experienced inflation (a broad and persistent rise in prices)

T10YIE and T5YIE from 10-Year & 5-Year Breakeven Inflation Rate | FRED | St. Louis Fed (fred.stlouisfed.org)

SPCC Presentation
June 8, 2021

15

Hilltop Economics

Highly Optimistic Consensus Semiconductor Outlook

- Pandemic-driven behaviors boosted demand in key semiconductor markets
- Strong final demand forecast and strong chip demand lead to optimism in the semiconductor forecasting community
- Revenue Outlook Strong
  - Three years WELL above trend
  - No pessimistic forecasters
  - MSI Wafers: +10-12% in 21

07-19: 5.1%/Yr	2020	2021	2022
Consensus (N=11)	**6.6**	**14.8**	**9.1**
Highest		20.0	12.0
Lowest		8.0	3.4

SPCC Presentation
June 8, 2021

16

Hilltop Economics

U.S. Spending on Tech Goods Accelerated When the Pandemic Hit

Spending momentum had been slowing as of 20Q1, but reached a new peak 21Q1

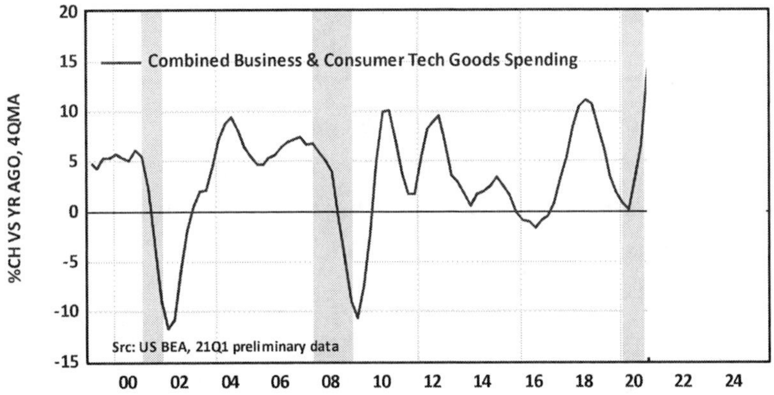

Shaded areas NBER U.S. recessions, end of current recession estimated Hilltop Economics

SPCC Presentation
June 8, 2021

Hilltop Economics

U.S. Spending on Tech Goods Accelerated When the Pandemic Hit

Spending momentum had been slowing as of 20Q1, but reached a new peak 21Q1

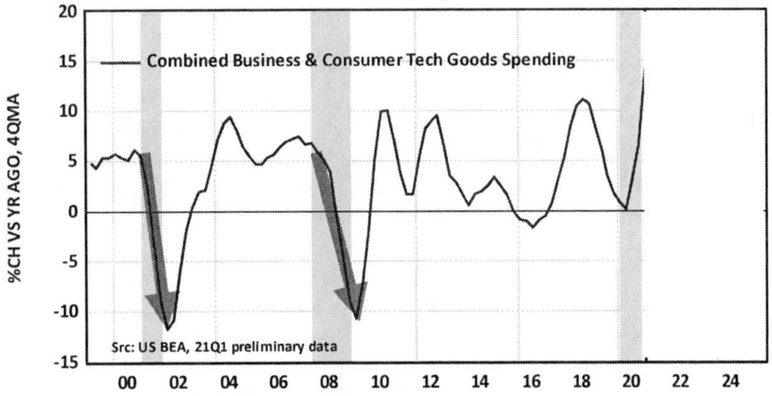

Shaded areas NBER U.S. recessions, end of current recession estimated Hilltop Economics

SPCC Presentation
June 8, 2021

Hilltop Economics

The Pandemic Was NOT a "Normal" Recession for Tech

- Tech During the Pandemic:
  - E-commerce & delivery apps allowed people to avoid all in-store shopping
  - Videoconferencing and collaboration apps enabled remote work
  - Videoconferencing and collaboration apps enabled online schooling
  - Videoconferencing and collaboration apps enabled remote medicine
  - Streaming services and video games provided in-home entertainment.
  - Videoconferencing enabled remote socializing
  - E-commerce & delivery apps allowed people to dine "out" in the home
- Tech goods surged as many tech goods became necessities, not the discretionary items people could cut back on when incomes fell.

SPCC Presentation
June 8, 2021

Hilltop Economics

U.S. Spending on Tech Goods Accelerated When the Pandemic Hit

Spending momentum had been slowing as of 20Q1, but reached a new peak 21Q1

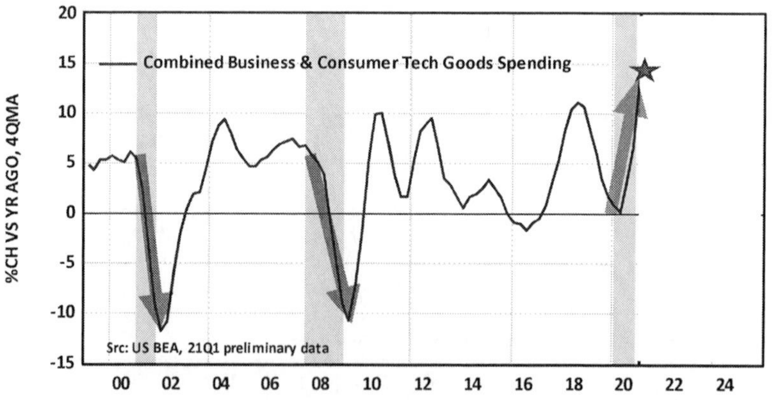

Shaded areas NBER U.S. recessions, end of current recession estimated Hilltop Economics

SPCC Presentation
June 8, 2021

Hilltop Economics

What Changes After the Pandemic?

Here is where the disagreement begins

- Remote everything fades
  - In-person work -> partly recovers
  - In-person education -> fully recovers, at least K-12
  - In-store retail -> some recovery
  - In-person entertainment -> some recovery
  - In-person socializing -> full recovery
  - In-person dining "out" in home -> full recovery
  - Travel -> full recovery leisure; partial business
- McKinsey studies:
  - Chinese consumers acting/spending as they did in pre-crisis times, EXCEPT for international travel - Charting the path to the next normal, Feb 3, 2021
  - Digital Adoption in Europe & US SLIPPING from Pandemic levels as of 6 months-ending April 2021 (still above prepandemic levels) – McKinsey Week in Charts 6/5/2021
- RISK: many tech good demand forecasts "trend" off current peak growth rates...
- AND CURRENT PEAKS ARE HIGH as of 2021Q1

SPCC Presentation
June 8, 2021

Hilltop Economics

Spending on Tech Goods Accelerated When the Pandemic Hit

Spending momentum had been slowing as of 20Q1, but reached a new peak 21Q1

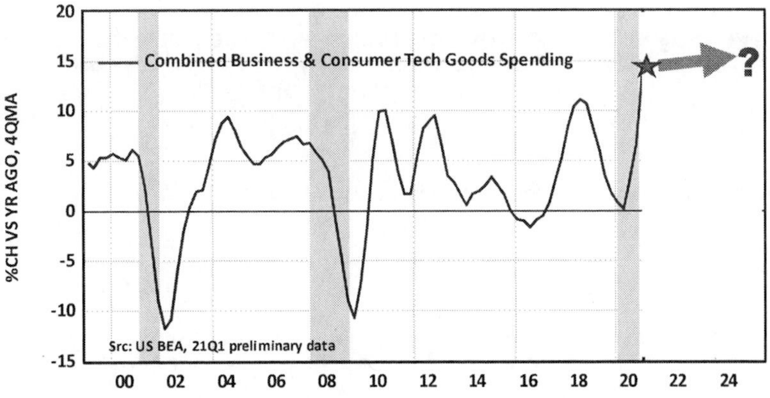

Shaded areas NBER U.S. recessions, end of current recession estimated Hilltop Economics

SPCC Presentation
June 8, 2021

Hilltop Economics

Spending on Tech Goods Accelerated When the Pandemic Hit

Spending momentum had been slowing as of 20Q1, but reached a new peak 21Q1

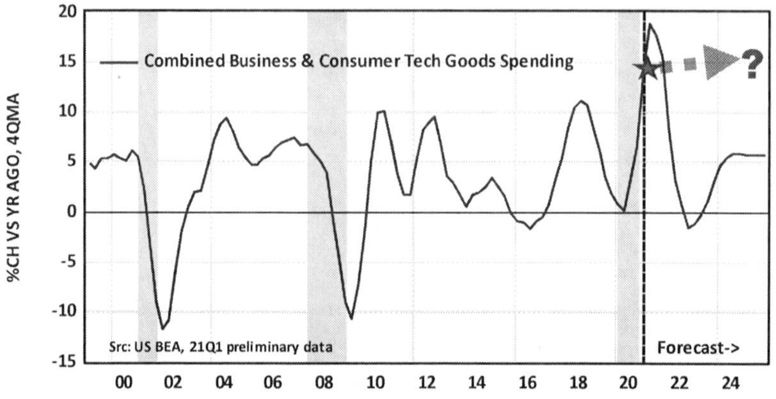

Shaded areas NBER U.S. recessions, end of current recession estimated Hilltop Economics

SPCC Presentation
June 8, 2021

Hilltop Economics

The Consensus Outlook Gives A Very Healthy Economic Backdrop For Semiconductors

	Share of Real GDP				Trendline Growth %/YR	
	2019	2020	2021	2025	2007-2019	2019-2025
Real GDP	100	100	100	100	2.5	2.9
Consumption	57.3	56.2	56.5	56.6	2.4	2.9
Investment	26.1	25.9	26.2	27.1	3.4	3.6
Government + Netex	16.6	17.9	17.4	16.3	1.4	2.2
Semi Wafer MSI					*4.0*	*7.4*

SPCC Presentation
June 8, 2021

24

Hilltop Economics

External Factors That Could Disrupt Semiconductors

- Downside risks
  - Post-pandemic spending pattern shifts away from tech goods even more than expected
  - Excess capacity develops – excess optimism and/or policy
  - Cross-border trade issues
- Upside risks
  - Post-pandemic economy retains pandemic tech-spending pattern
  - 5G investment accelerates, new technological advances from international competition for capacity

Hilltop Economics

Summary Overview

- **Economic Recovery Underway**
  - **Fueled by stimulus packages**
  - **Inflation the primary risk**
- **Consensus Outlook Continues to Rise**
  - **Continued government support**
  - **Strong consumer demand**
  - **Strong business demand**
- **Upstream Semiconductor Markets Remain Strong**
  - **Anticipate some softening once the pandemic recession fades.**

Hilltop Economics

Duncan Meldrum
Chief Economist
Hilltop Economics, LLC
duncan.meldrum@hilltopeconomics.com

610.709.7951

Challenges in next generation semiconductor manufacturing

Ph.D Kuntack Lee
Process Development
Semiconductor R&D Center
Samsung Electronics. Co

Ph.D Kuntack Lee (李 根 澤)

VP of Technology
Process Development
Semiconductor R&D Center
SAMSUNG Electronics

[Education]

. **BS in Chemistry, Seoul National University**
. **Ph.D in Physical Chemistry, POSTECH**

[Work Experience]

. **2020 Program Committee, SPCC**
. **2015 VP of Technology**
. **2014 Samsung Award of Honor**
. **2010 Program Committee, UCPSS**
. **1996 Samsung Electronics Co., Ltd**

e-mail : kuntack.lee@samsung.com

Semiconductor Manufacturing

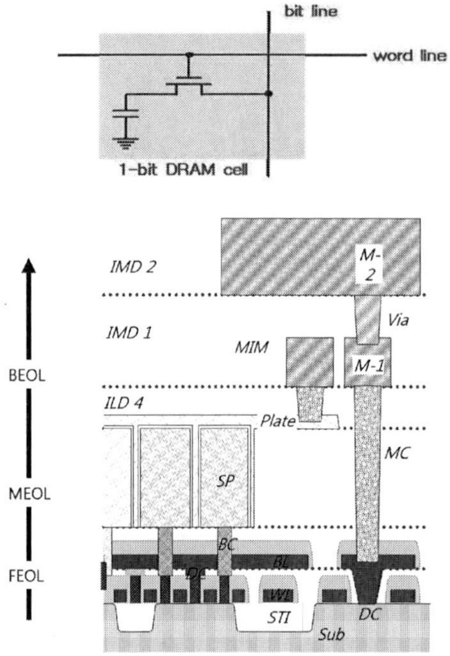

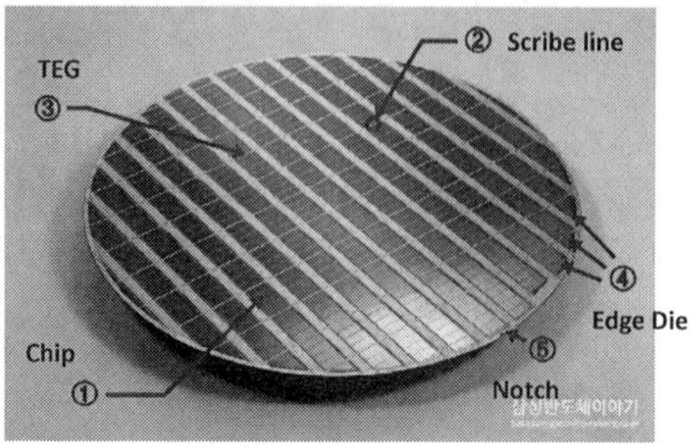

Cooking by using Equipment & Materials

+

Gas, Chemical, Energy

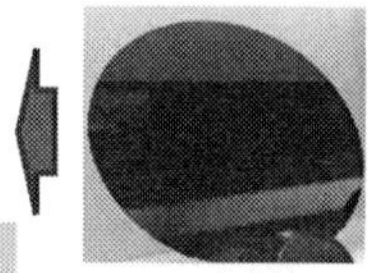

Process, Equipment & Infra

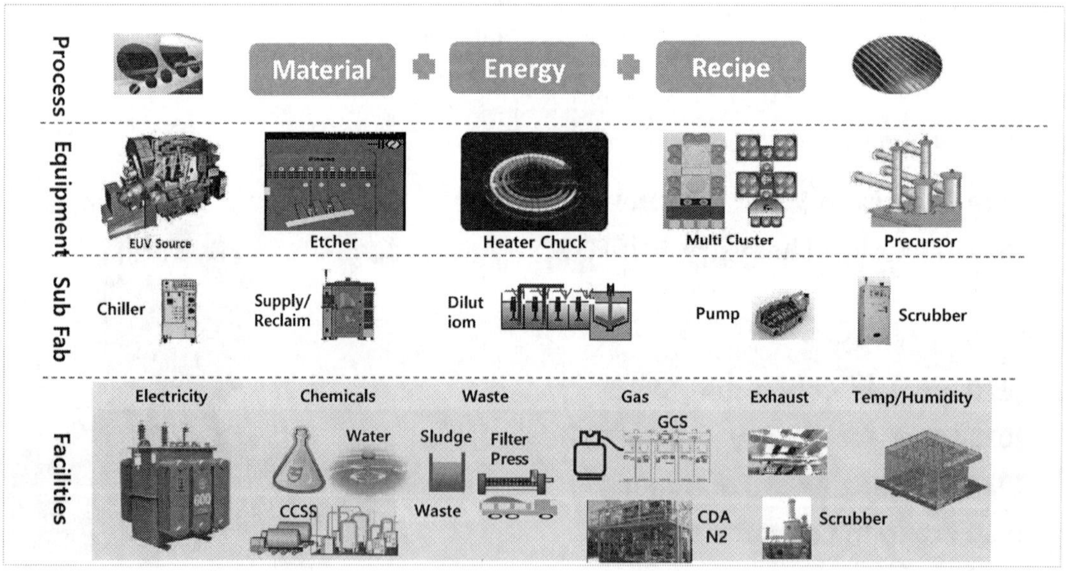

New Tech & Infra Consumption

Ultra Fine Patterning	J¤ý¯ !vζ'∝ψx-ψΣ	High Conductance

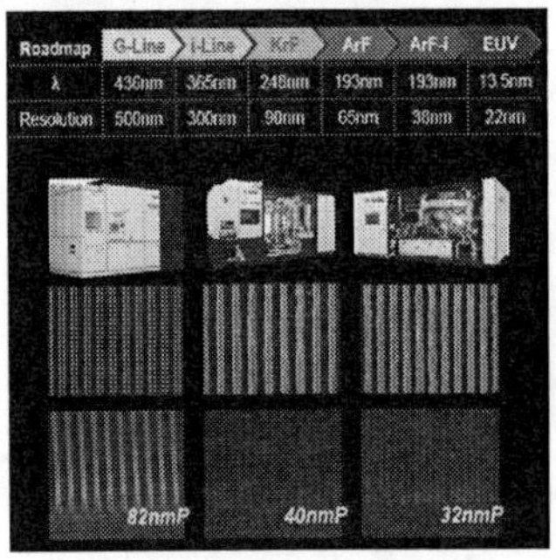

◆ EUV: Power Consumption ↑

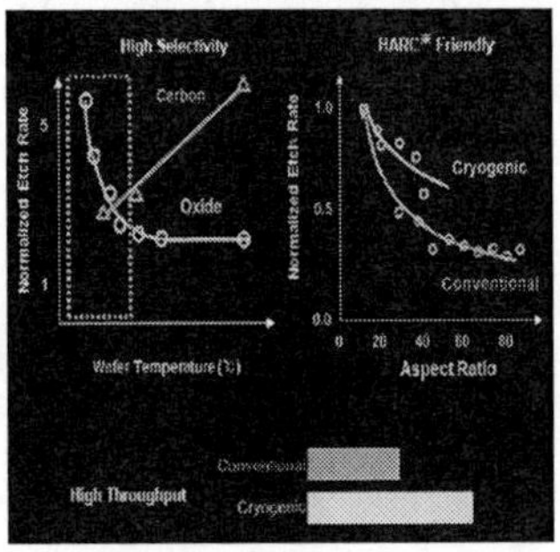

◆ Cryo Etch: Chiller Area ↑

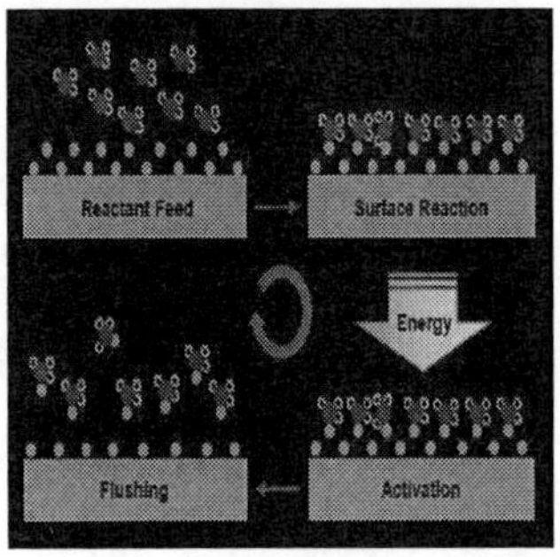

◆ ALD : Pump Capacity ↑

New Technologies require more power and larger area.

Power Consumption by Facility & Sub Fab

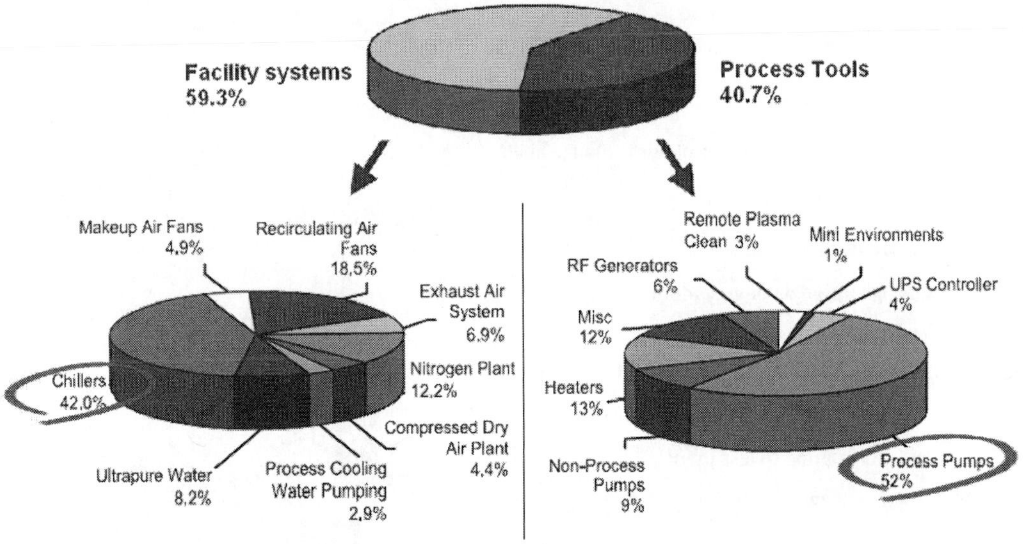

Pumps and Chillers are the most consumer of power.

Power Consumption by pump & chiller

Pump Capacity vs Area

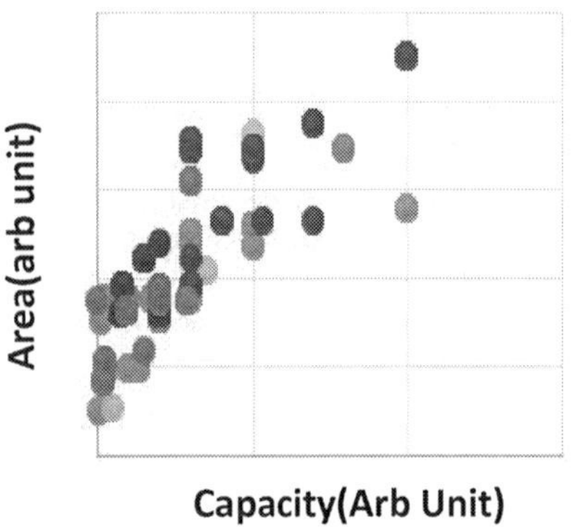

Chiller Capacity Trend

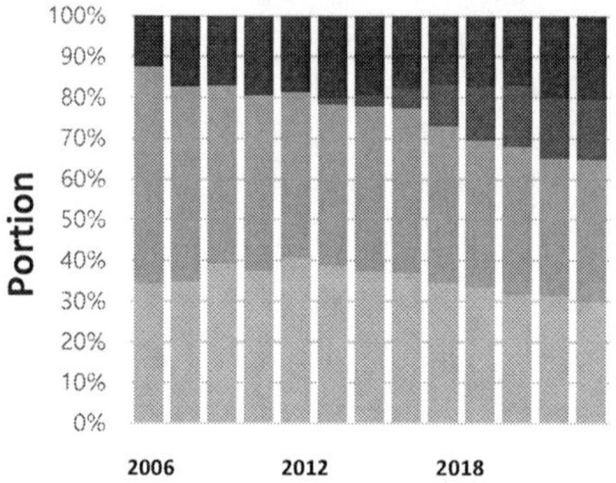

Capacity and area are continuously increasing.

Power Consumption Reduction by synchronization

- The Green Factory concept is important for the future
- To measure the energy consumption level for all tools
  - This will be reflected in the tool performance ability.

- In Equipment
  - Tool Design based on energy reduction
  - Increased communication between tool and sub tools
  - In Idle mode, each tool needs to reduce Energy Consumption to less than 50% of Process mode.

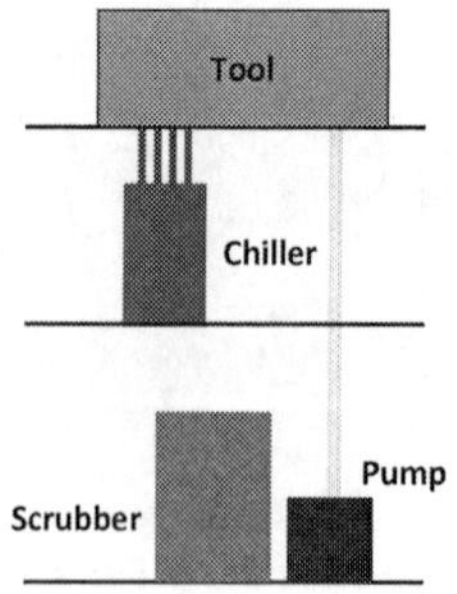

Synchronization!

Process	Idle
↓	↓
Process	Idle
↓	↓
Process	Idle

Sub tool synchronization is needed.

Water Consumption

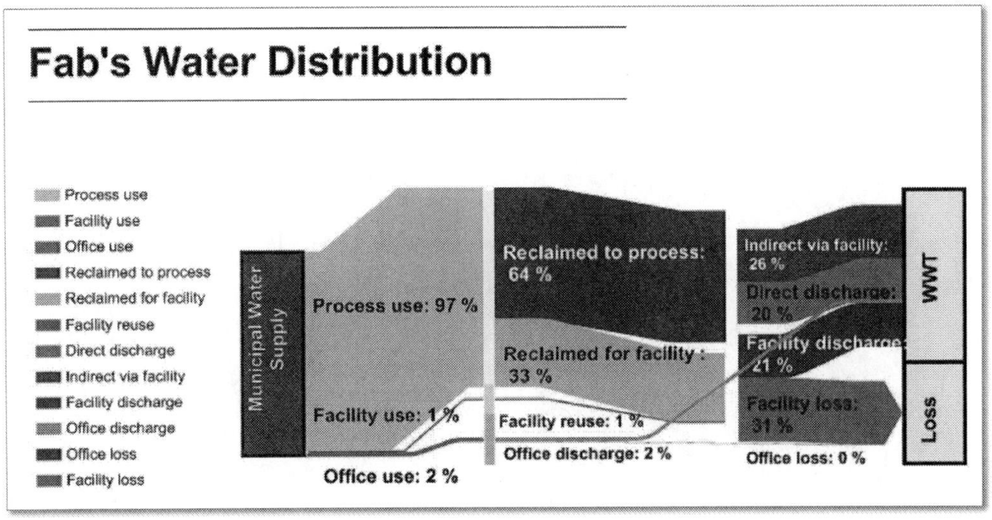

Most water are reused and dry processes are implemented.

Chemical Consumption

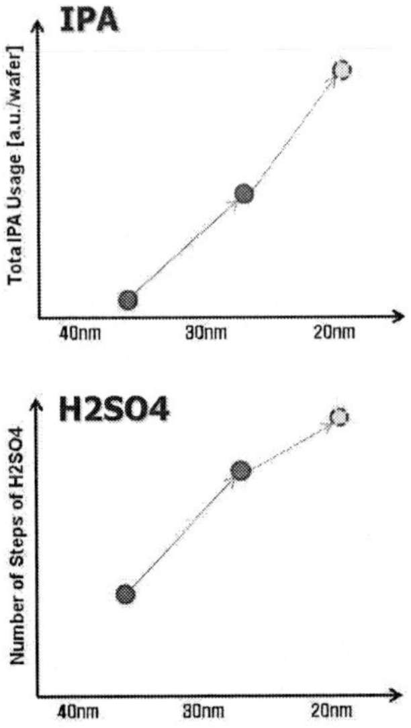

- Increase in IPA Usage>
  - Batch Cleaner → Single Wafer Cleaner
  - Design Rule Shrink → Vulnerable to Leaning
- Increase in H_2SO_4 Consumption Amount/Steps >
  - Patterning: SET → Double/Quadruple Patterning Tech.
  - Batch Cleaner → Single Wafer Cleaner
  - Increment in Integration Steps

: Reduction (Replace with other process, Recipe change)
Reuse & Recycle

IPA and Sulfuric acid usages are increasing continuously.

Waste Reduction by Reuse & Recycle

[Slurry]

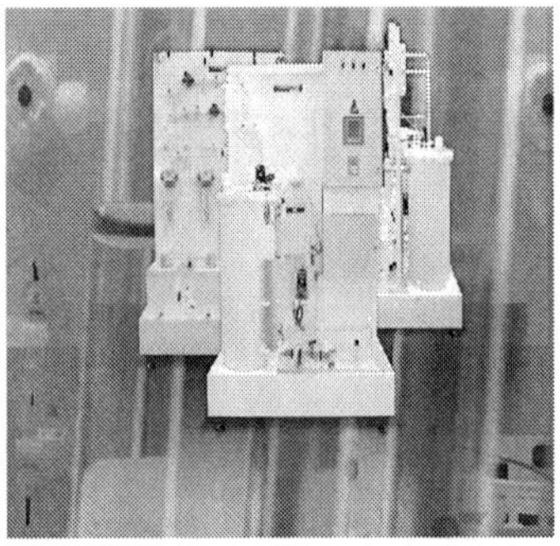

[HF]

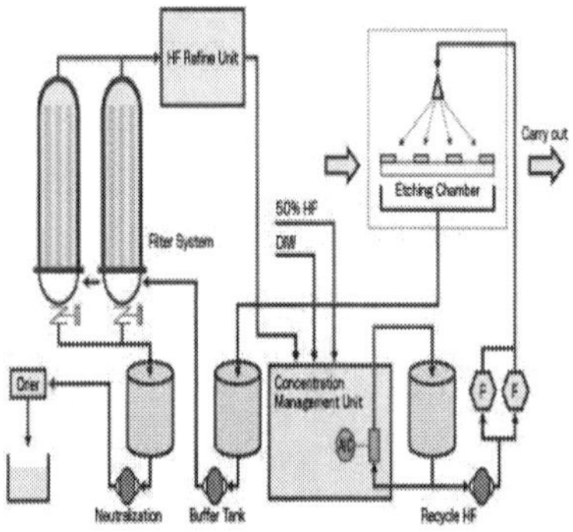

CMP slurry and HF Chemical recycle systems are developed.

Metal Gas Recycle

WF6 Exhaust System

Main Tool

Vacuum pump

By product
:HF, WFx, BHx, WOFx

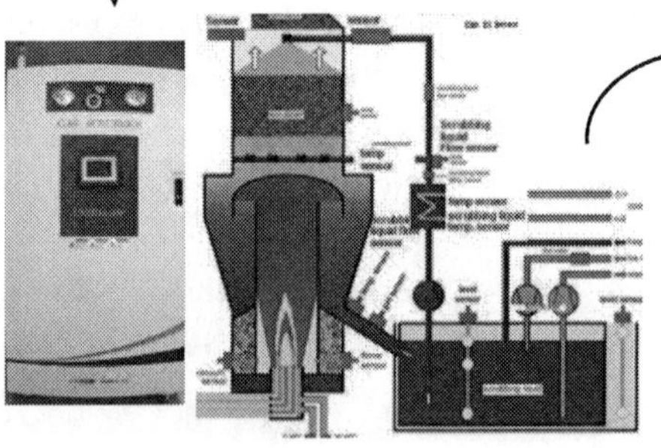

Scrubber (Burn Wet)

WOx in water

Recycle ratio is too low!

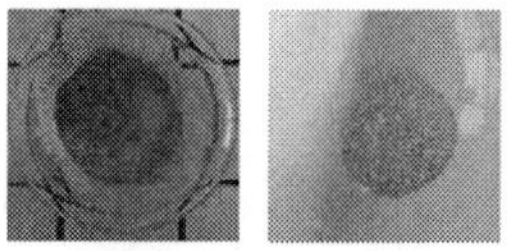

WOx·HyO WOx

W metal recycle system has been tried but not economic yet.

Gas Recycle Systems Developed

[NF3]   [H2]   [Xe]

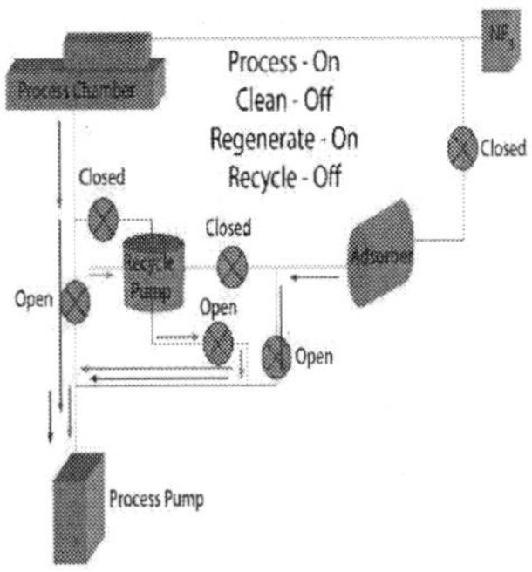

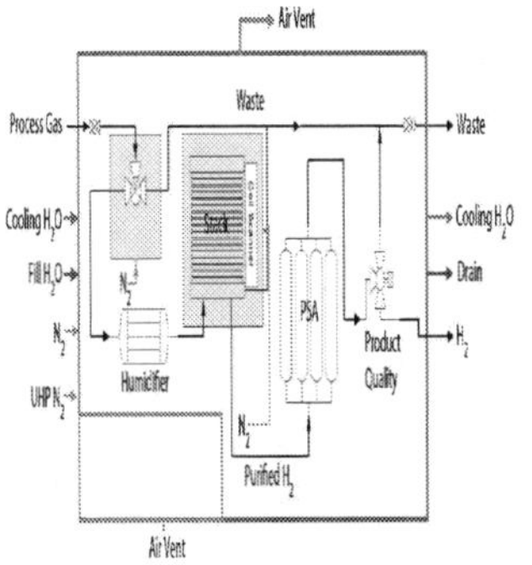

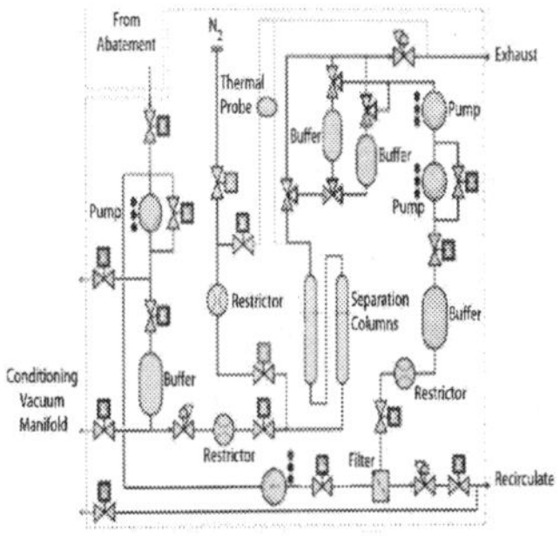

Recycle systems are developed for Bulk & Process gases.

Replace long time etch with wet etch (AAO)

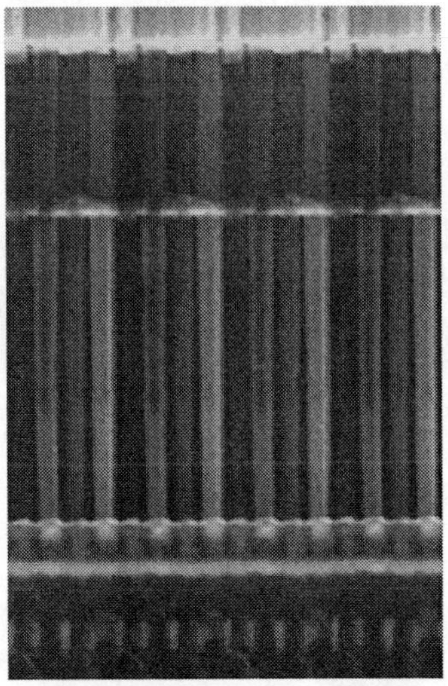

From Google Image

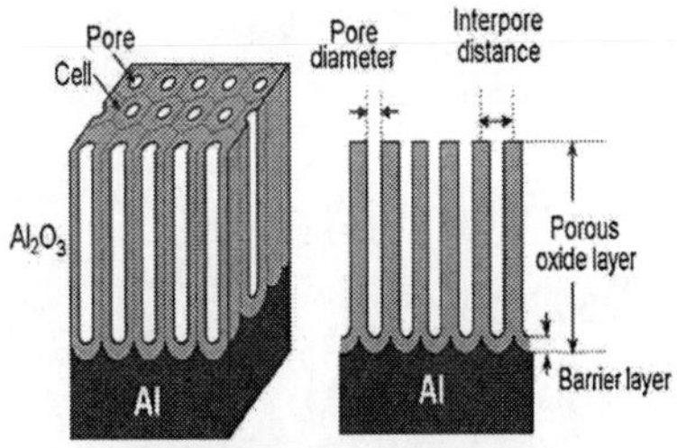

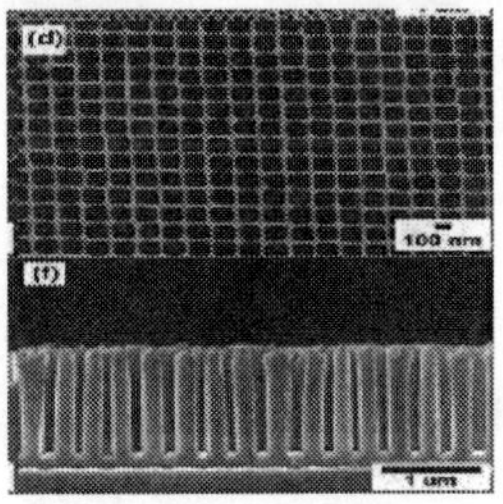

Is it possible to replace dry etching with wet etching?

AAO Deep Contact Formation Method

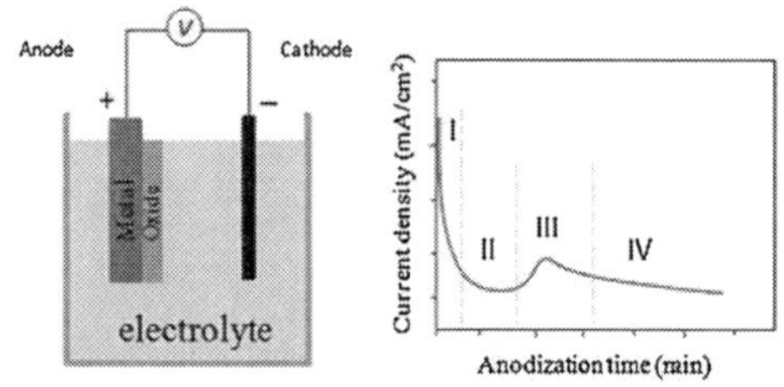

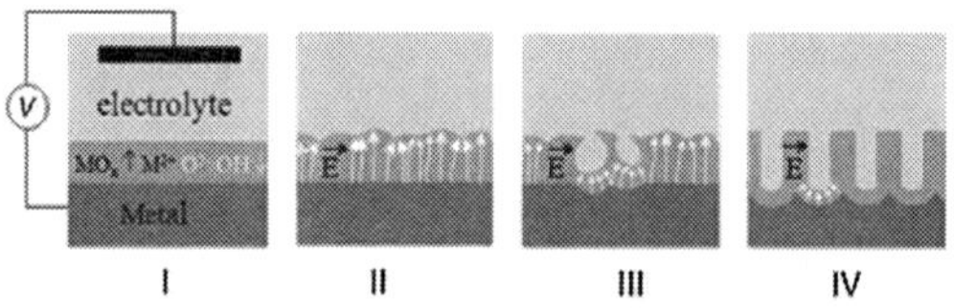

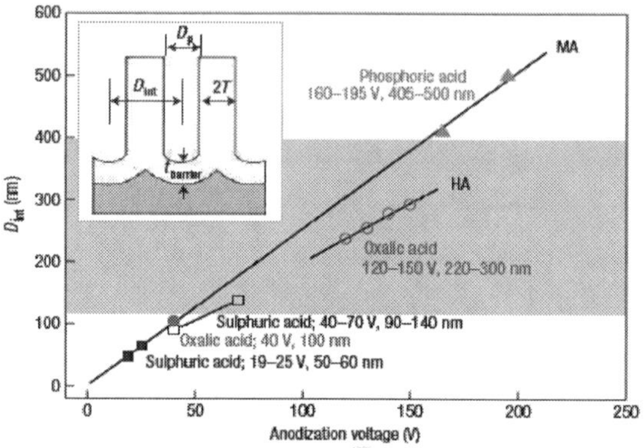

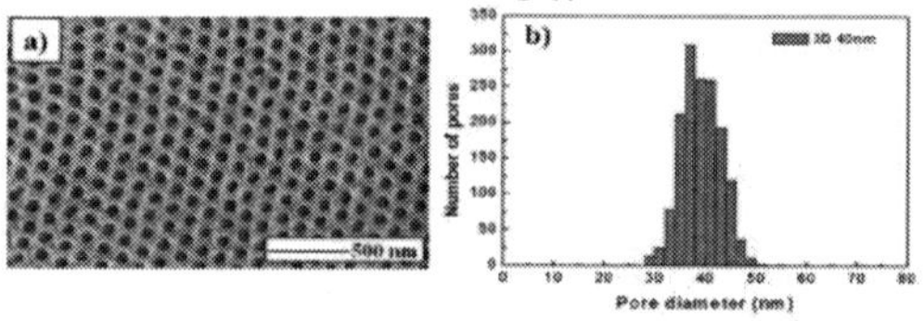

Pore size distribution and porosity can be overcome ?

Same Challenges are still Here

2003, Semicon Europa

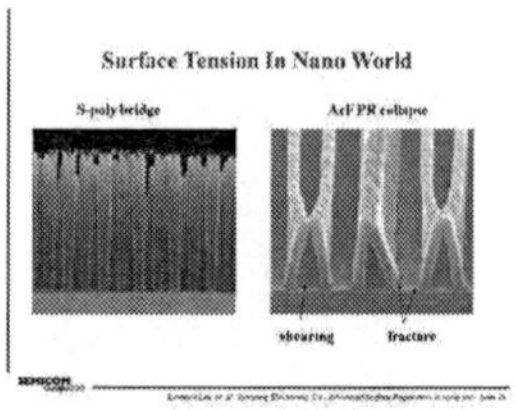

2001, SCP Symposium

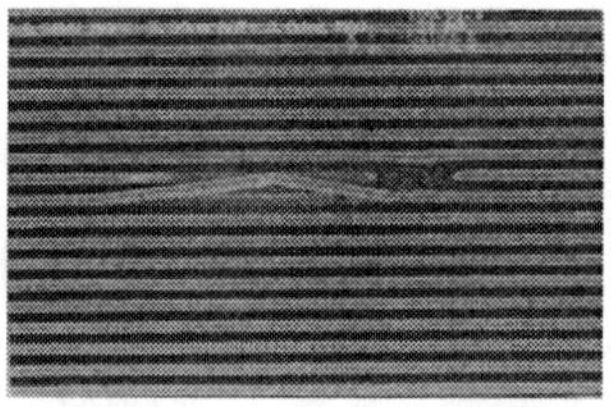

2001, SCP Symposium

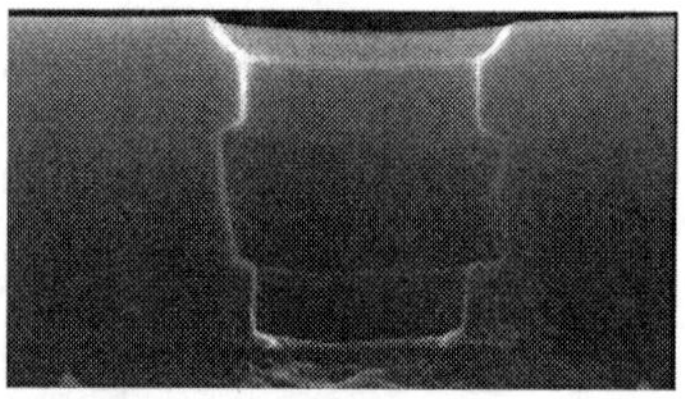

2012, Clean UGM

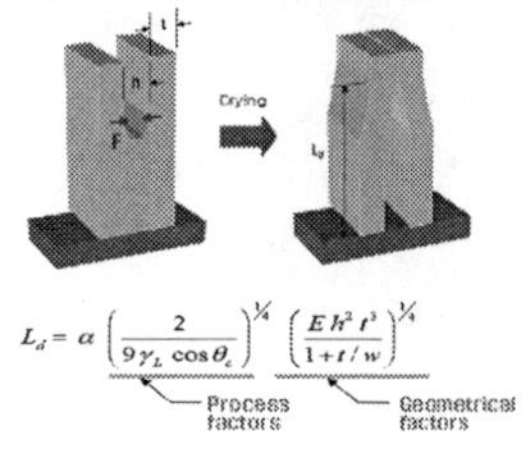

2008, Semicon West

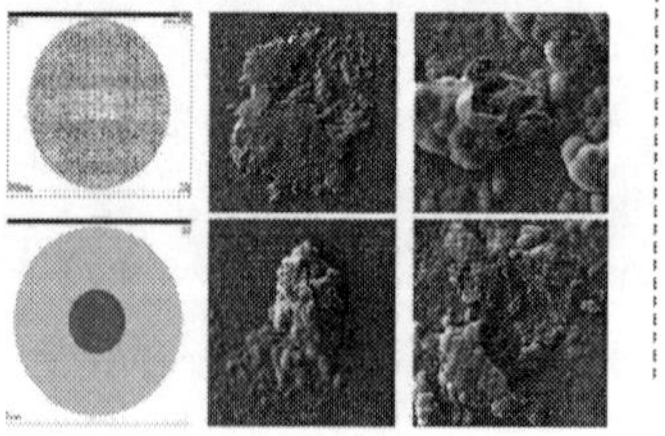

2012, Clean UGM

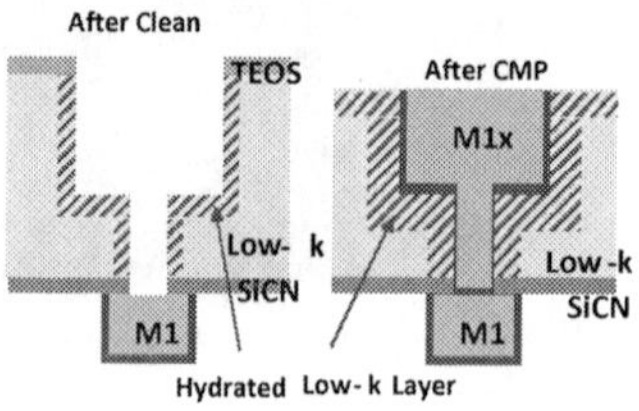

Drying for Deeper & Narrower | **Smaller Particle on Narrower Line** | **Porous & Corrosive Material clean**

Summary

- New Technology requires
  more energy and larger area.

- Power consumption by sub tools
  should be reduced.

- Chemical usage should be reduced by
  reuse and recycle.

- Can the new process
  replace current gas consuming process?

- Same kind of Cleaning challenges are still here.

SPCC 2021, June 9th   17/17   SAMSUNG

Chemical Quality in Advanced Fab Processes
intel.
Intel Unleashed
Engineering the Future
June 8, 2021 - SPCC Business of Cleans Conference
Jeff Hemphill, Global Supply Chain Fab Materials Organization

Our purpose is to create world-changing technology that enriches the lives of every person on earth.
Intel Unleashed
Engineering the Future
intel

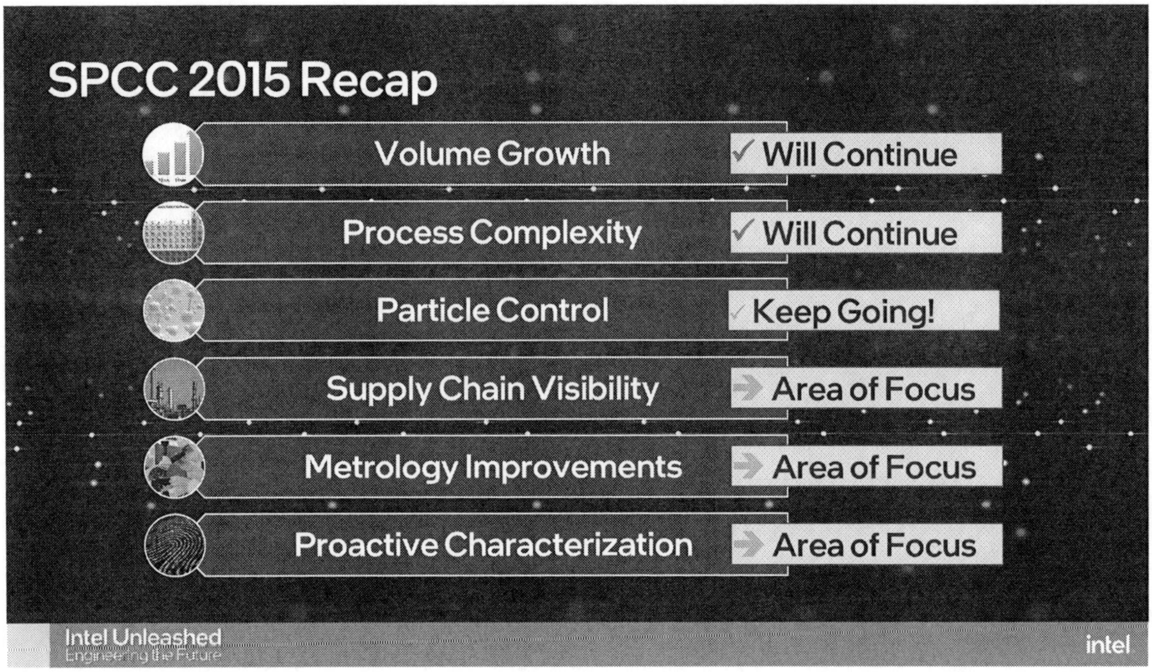
SPCC 2015 Recap
Volume Growth
✓ Will Continue
Process Complexity
✓ Will Continue
Particle Control
Keep Going!
Supply Chain Visibility
Area of Focus
Metrology Improvements
Area of Focus
Proactive Characterization
Area of Focus
Intel Unleashed
Engineering the Future
intel

Process Chemical Quality for 2021+
Cost of Quality
Manufacturing Excellence
Design for Quality
Intel Unleashed
Engineering the Future
intel

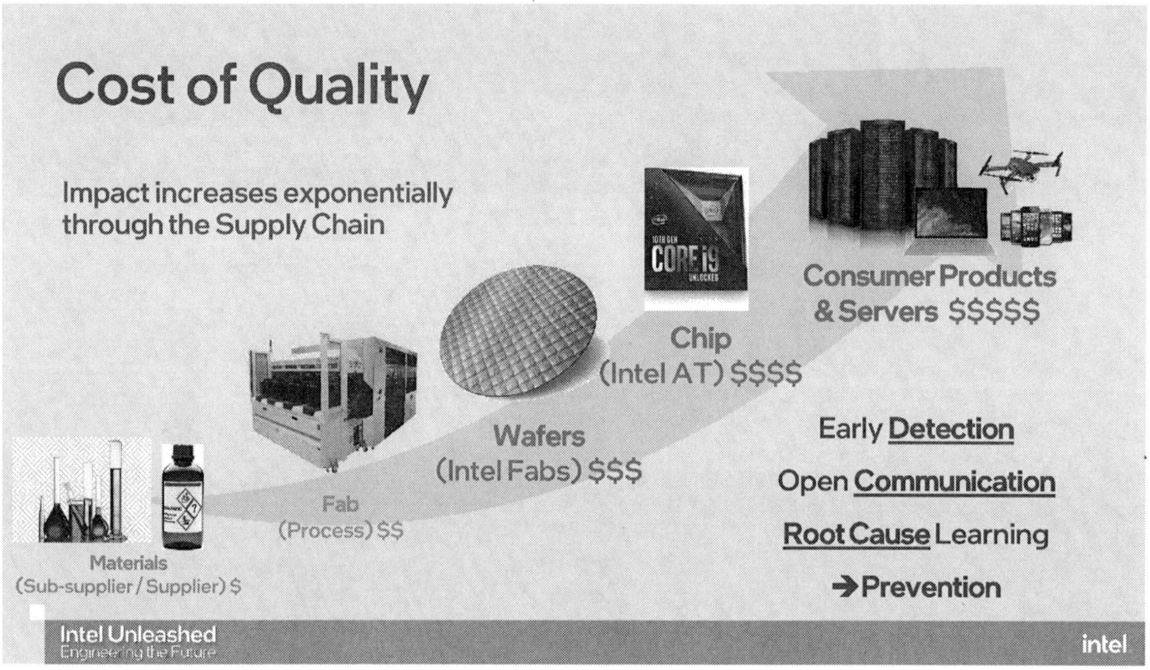
Cost of Quality
Impact increases exponentially through the Supply Chain
Consumer Products & Servers $$$$$
Chip (Intel AT) $$$$
Wafers (Intel Fabs) $$$
Fab (Process) $$
Materials (Sub-supplier / Supplier) $
Early Detection
Open Communication
Root Cause Learning
→Prevention
Intel Unleashed
Engineering the Future
intel

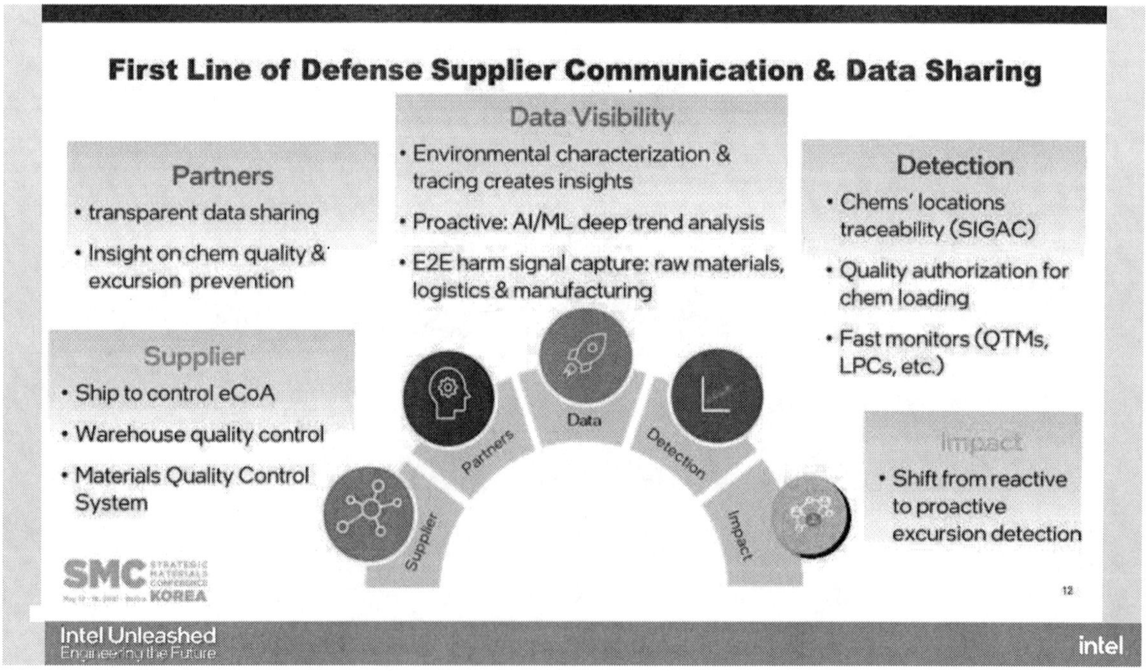
First Line of Defense Supplier Communication & Data Sharing
Data Visibility
• Environmental characterization & tracing creates insights
• Proactive: AI/ML deep trend analysis
• E2E harm signal capture: raw materials, logistics & manufacturing
Partners
• transparent data sharing
• Insight on chem quality & excursion prevention
Detection
• Chems' locations traceability (SIGAC)
• Quality authorization for chem loading
• Fast monitors (QTMs, LPCs, etc.)
Supplier
• Ship to control eCoA
• Warehouse quality control
• Materials Quality Control System
Impact
• Shift from reactive to proactive excursion detection
Supplier
Partners
Data
Detection
Impact
SMC STRATEGIC MATERIALS CONFERENCE KOREA
12
Intel Unleashed
Engineering the Future
intel

Cost of Quality

Wet Cleans = Big Impact

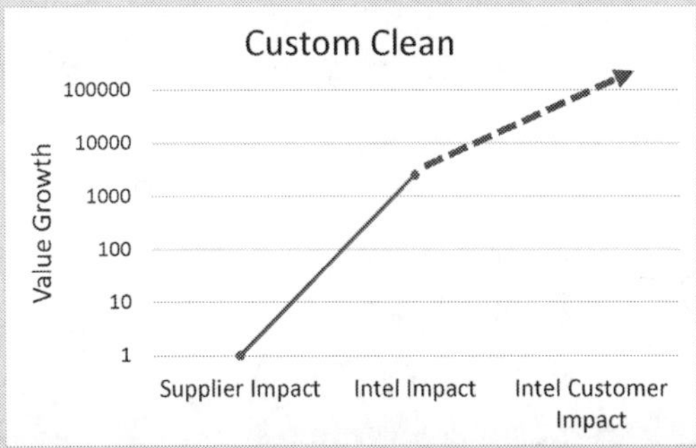

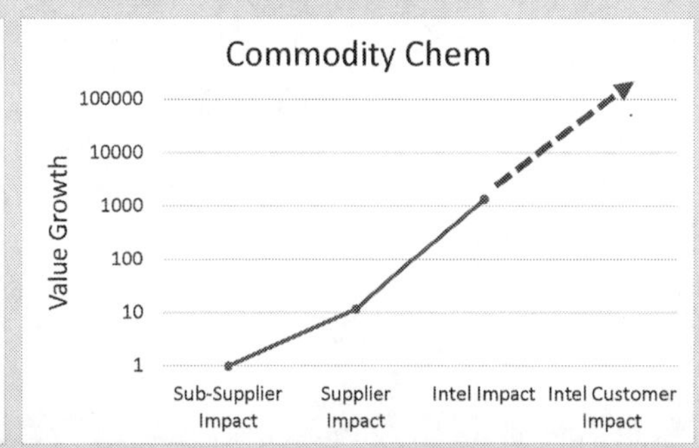

Material	Quality Issue	Sub-supplier Impact	Supplier Impact	Intel Impact
Custom Clean	Mfg. system contamination from operational cleanliness issue.	NA	1x	2,500x
Commodity Chem	Sub-Supplier low-grade chemical flush after maintenance	1	12x	1,350x

Intel Unleashed
Engineering the Future

Manufacturing Excellence

Embrace Cleanroom Culture

Pristine Manufacturing Environment

Robust Monitoring & Procedures

Quality is a Value

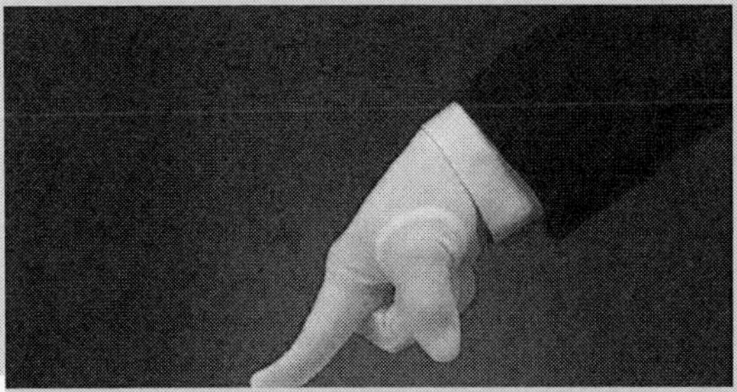

Intel Unleashed
Engineering the Future

Manufacturing Excellence

Quality Supply Chain

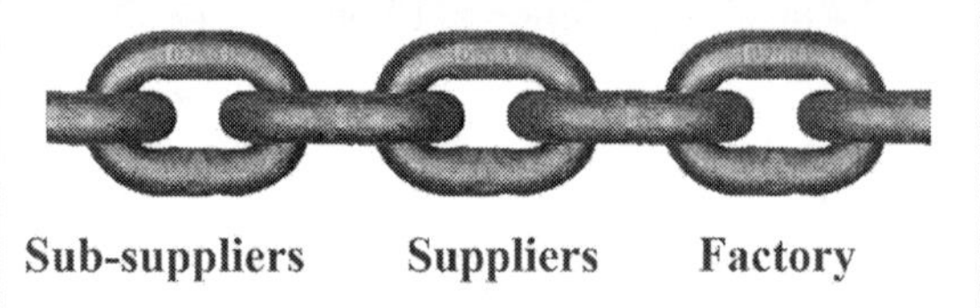

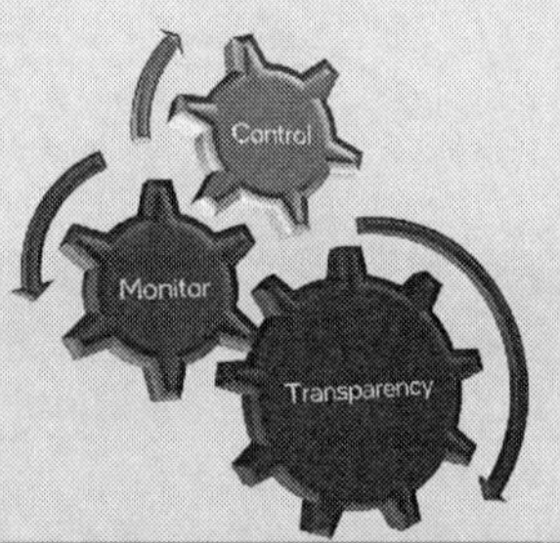

Holistic Chemical Manufacturing Process:

Raw Materials

Packages & Filters

Shipping & Warehousing

Labor & Training

Facilities / UPW Systems

Monitoring

Intel Unleashed
Engineering the Future

intel

Manufacturing Excellence

Continuous Improvement

Sensitivity Increasing

Follow Technology Roadmaps

Be Proactive and Collaborate

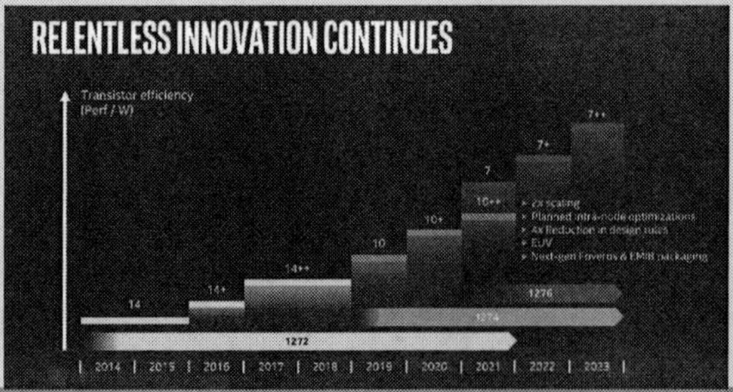

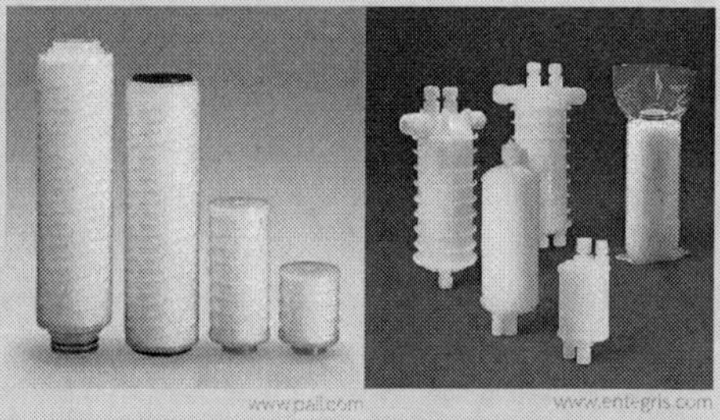

Intel Unleashed
Engineering the Future

intel

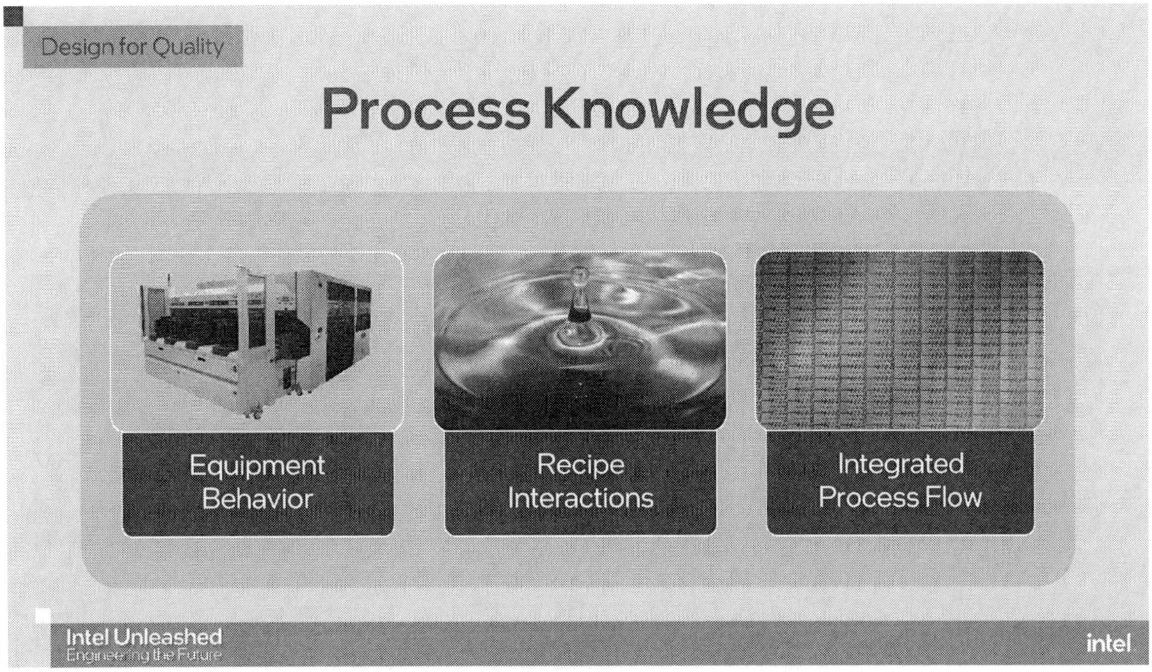
Design for Quality
Process Knowledge
Equipment Behavior
Recipe Interactions
Integrated Process Flow
Intel Unleashed
Engineering the Future
intel

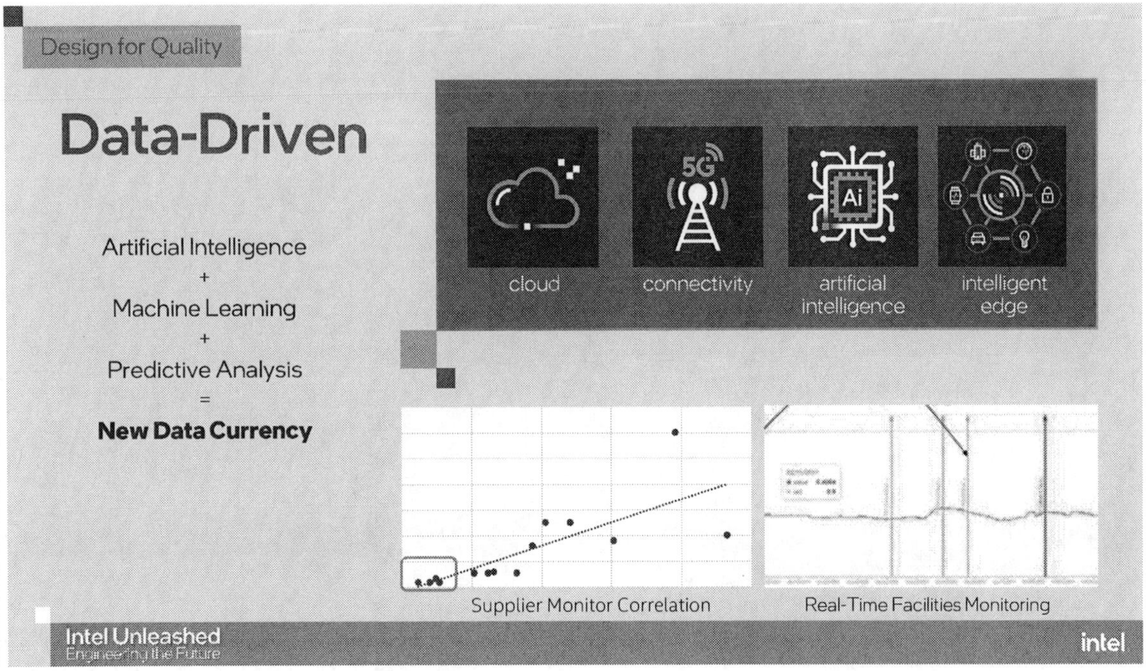
Design for Quality
Data-Driven
Artificial Intelligence
+
Machine Learning
+
Predictive Analysis
=
New Data Currency
5G
Ai
cloud
connectivity
artificial intelligence
intelligent edge
Supplier Monitor Correlation
Real-Time Facilities Monitoring
Intel Unleashed
Engineering the Future
intel

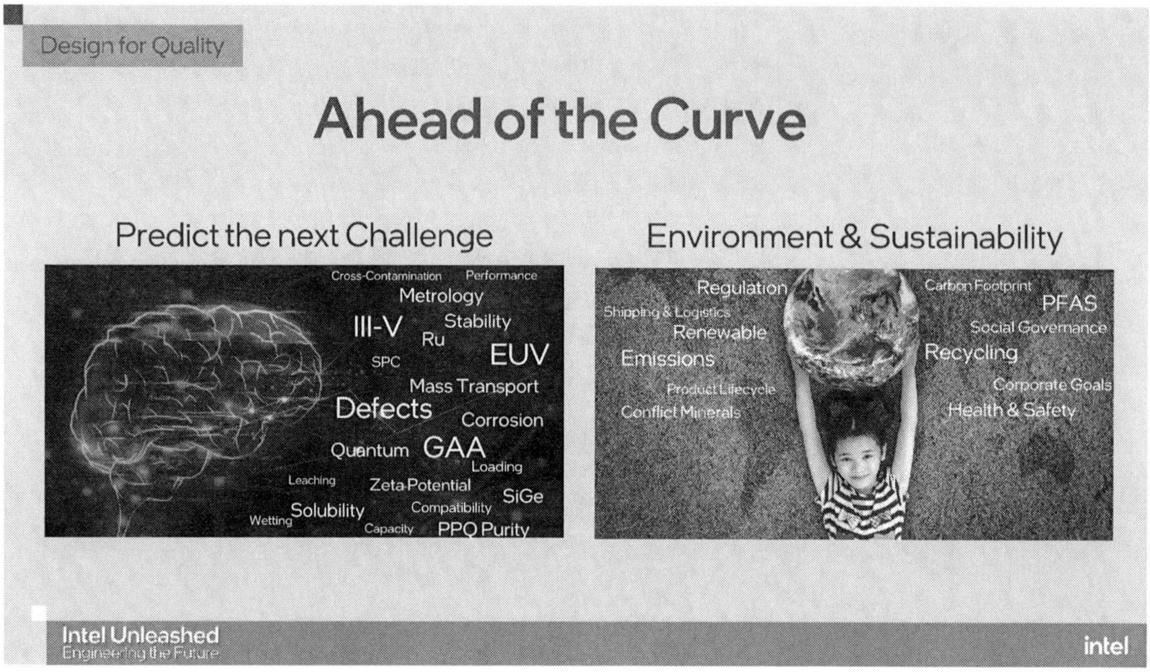
Design for Quality
Ahead of the Curve
Predict the next Challenge
Cross-Contamination
Performance
Metrology
III-V
Stability
Ru
SPC
EUV
Mass Transport
Defects
Corrosion
Quantum
GAA
Loading
Leaching
Zeta Potential
SiGe
Wetting
Solubility
Compatibility
Capacity
PPQ Purity
Environment & Sustainability
Regulation
Carbon Footprint
PFAS
Shipping & Logistics
Renewable
Social Governance
Emissions
Recycling
Product Lifecycle
Corporate Goals
Conflict Minerals
Health & Safety
Intel Unleashed
Engineering the Future
intel

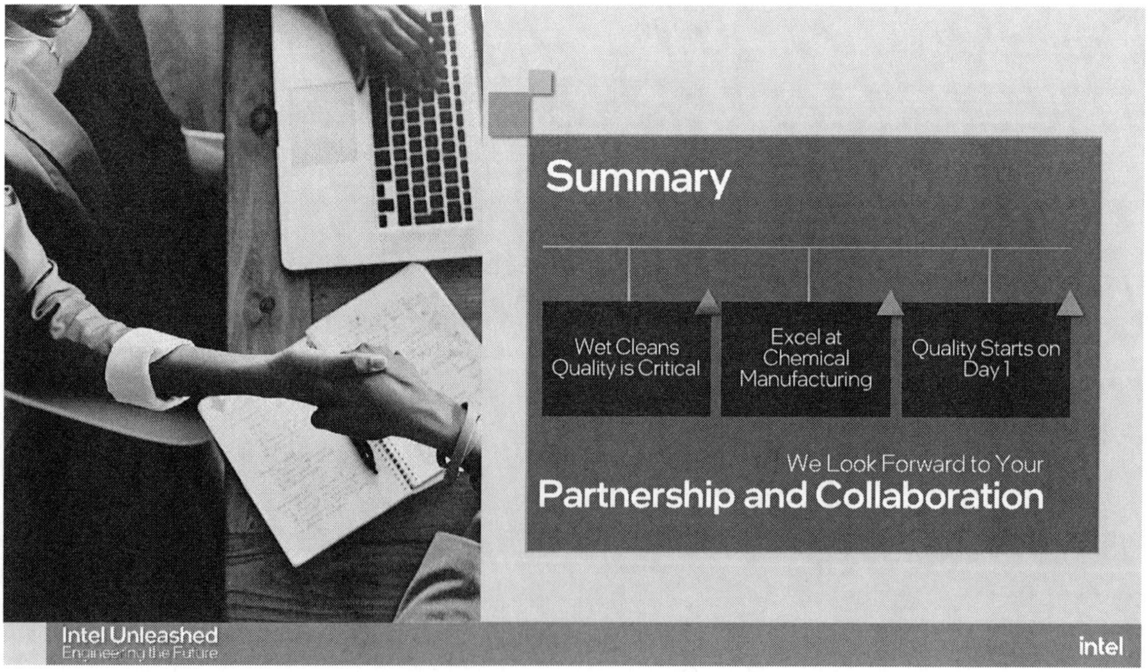
Summary
Wet Cleans Quality is Critical
Excel at Chemical Manufacturing
Quality Starts on Day 1
We Look Forward to Your
Partnership and Collaboration
Intel Unleashed
Engineering the Future
intel

Notices & Disclaimers

Intel does not control or audit third-party data. You should consult other sources to evaluate accuracy.

All product plans and roadmaps are subject to change without notice.

Statements in this document that refer to future plans or expectations are forward-looking statements. These statements are based on current expectations and involve many risks and uncertainties that could cause actual results to differ materially from those expressed or implied in such statements. For more information on the factors that could cause actual results to differ materially, see our most recent earnings release and SEC filings at www.intc.com.

Any forecasts of goods and services needed for Intel's operations are provided for discussion purposes only. Intel will have no liability to make any purchase in connection with forecasts published in this document.

Intel Unleashed
Engineering the Future

intel.

Wet Processes at the Leading Edge

Scotten W. Jones – President – IC Knowledge LLC
sjones@icknowledge.com

IC KNOWLEDGE LLC

Outline

IC KNOWLEDGE LLC

- High level overview of wet processes.
  - Wet cleans, etches and strips.
- For three key industry segments discuss the technology and technology trends in the segment, and then the impact on wet cleans:
  - 3D NAND
  - DRAM
  - Logic
- Conclusion

3D NAND Bit Density Trend

IC KNOWLEDGE LLC

- The transition from 2D NAND to 3D is enabling the continuation in bit density scaling by using the third dimension.
- Bit density is the number of gigabits of memory on the die divided by the die size.
- Multiple points for the same company in the same year represent MLC/TLC/QLC.

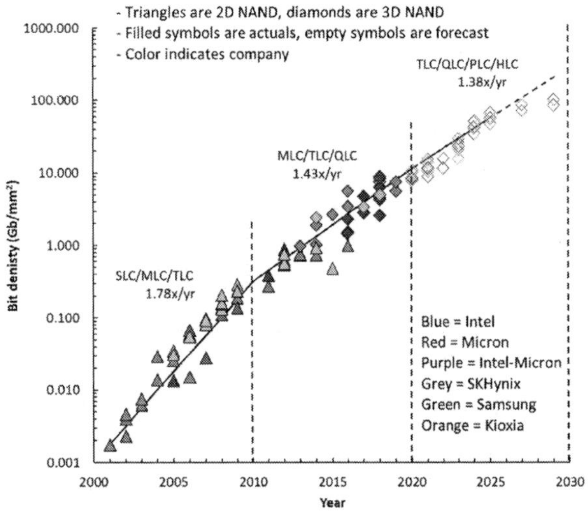

IC Knowledge and TechInsights data

3D NAND "Nodes"

IC KNOWLEDGE LLC

	2016	2017	2018	2019	2020	2021	2022	2023	2024	2025
Intel-Micron (FG)	**32L** 1 x 32 CUA	**64L** 2 x 32 CUA	**96L** 2 x 48 CUA	NA	NA	NA	NA	NA	NA	NA
Intel/ SKH (FG)	NA	NA	NA	NA	**144L** Int 3 x 48 CUA		192L SKH 4 x 48 CUA	240L SKH 5 x 48 CUA		288L SKH 6 x 48 CUA
Kioxia (CT)	**48L** 1 x 48 Side	**64L** 1 x 64 Side	**96L** 2 x 48 Side		**112L** 2 x 56 CUA	**162L** 2 x 81 CUA		243L 3 x 81 CUA	324L 4 x 81 CUA	512L 4 x 128 CUA
Micron (CT)	NA	NA	NA	NA	**128L** 2 x 64 CUA	**176L** 2 x 88 CUA		264L 3 x 88 CUA	352L 4 x 88 CUA	512L 4 x 128 CUA
SKH (CT)	**36L** 1 x 36 Side	**48L** 1 x 48 Side	**72L** 2 x 36 Side	**96L** 2 x 48 CUA	**128L** 2 x 64 CUA	**176L** 2 x 88 CUA		264L 3 x 88 CUA	352L 4 x 88 CUA	512L 4 x 128 CUA
SS (CT)	**64L** 1 x64 Side		**92L** 1 x 92 Side	**128L** 1 x 128 Side		160L 2 x 80 CUA		276L 3 x 92 CUA	368L 4 x 92 CUA	512L 4 x 128 CUA

3D NAND Scaling

IC KNOWLEDGE LLC

Layers (array only)

Layers	Strings	Scaling
32L	1 x 32	NA
48L	1 x 48	1.50
64L	1 x 64	1.33
92L	1 x 92	1.44
128L	1 x 128	1.39
160L	2 x 80	1.25
276L	3 x 92	1.73
368L	4 x 92	1.33
512L	4 x 128	1.39
640L	5 x 128	1.25
768L	6 x 128	1.20

Diminishing returns

Bits Per Cell

Type	Bits/cell	Scaling
SLC	1	NA
MLC	2	2.00
TLC	3	1.50
QLC	4	1.33
PLC	5	1.25

Diminishing returns

- Horizontal
  - Stair step -> folded stair case (<2%)
  - CMOS under array (~25%)
  - Tighter pitch, fewer slots (~5-10%)

3D NAND Process Challenges

IC KNOWLEDGE LLC

- Deposit alternating layers of SiO/SiN or SiO/pSi with good thickness control and low stress.
- Channel hard mask deposition and patterning.
- Etching and filling high aspect ratio (HAR) channel. Etch limit is ~48 pairs for SiO/pSi, ~128 pairs for SiO/SiN.
- High mobility pSi channel deposition and anneal, possible new materials in the future.
- Staircase formation – repeating etch, photoresist shrink and etch.
- For charge trap devices, strip SiN and refill with AlO/TiN/W with low resistance and stress. Possible new materials in the future.

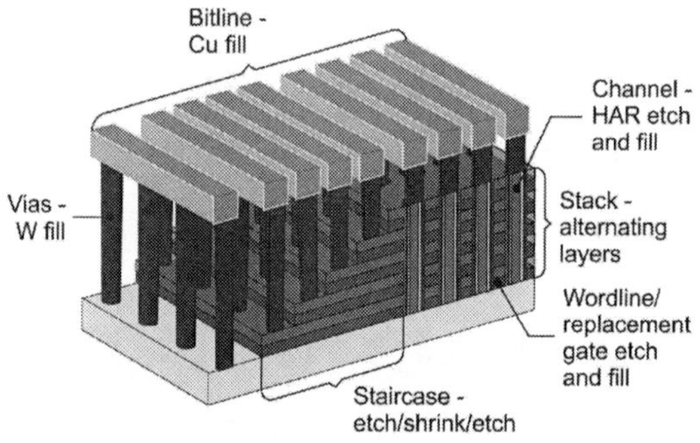

Charge Trap 3D NAND
(Kioxia, Micron, SK Hynix, Samsung)

Copyright IC Knowledge LLC, all rights reserved

3D NAND Wet Process Trends

IC KNOWLEDGE LLC

- Total wet process steps by layers
  - BEOL – post metal or via etch
  - BSB – backside bevel for litho or CMP
  - CMP – on-tool and post cleans
  - Etch – wet etches
  - FEOL – front end anhydrous HF, ammonium peroxide and critical cleans
  - SCCO2 – super critical CO2
  - Strip – photoresist and carbon strips
- Based on Samsung 3D NAND layers, string stacking and cleans usage.

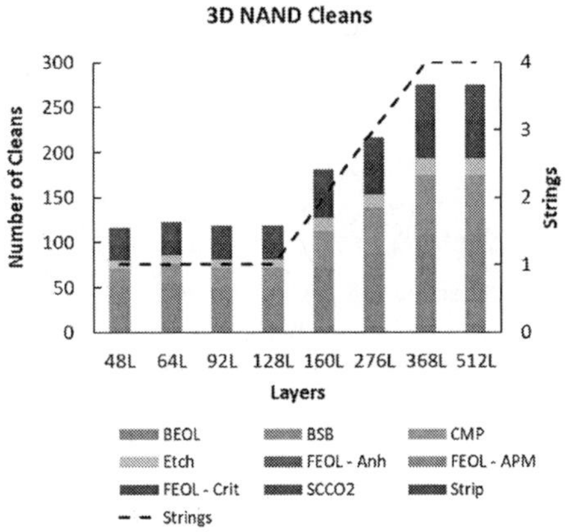

Source: IC Knowledge Strategic Cost and Price Model – 2021 – revision 00a

5/27/2021 7

DRAM Bit Density Trend

IC KNOWLEDGE LLC

- Bit density is die capacity in Gb divided by die size in mm².
- The solid black line is the long term trend based on actual values.
- The dashed black line is the forecasted trend going forward based on a mix of actual and forecast values.
- Bit density scaling began to slow around 2014/2015 primarily due to difficulties scaling the capacitor.

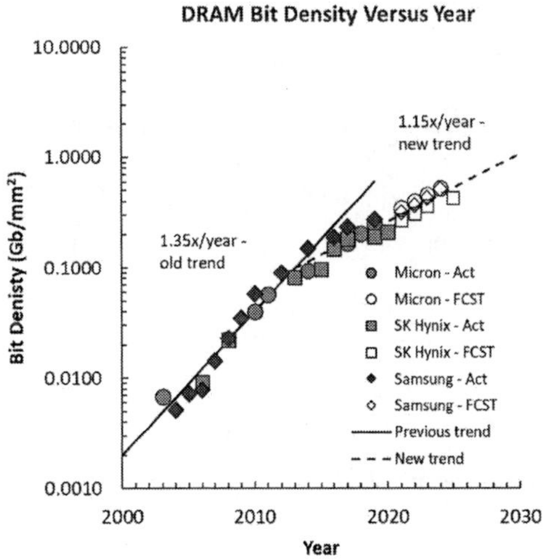

IC Knowledge and TechInsights data

5/27/2021 8

DRAM Nodes

IC KNOWLEDGE LLC

	2016	2017	2018	2019	2020	2021	2022	2023	2024	2025
MT		1x-19nm	1y-16nm	1z-15nm		1α – 13nm	1β - 12nm	1γ - 11nm	1δ - 10nm	
SKH		1x – 17nm	1Y – 16nm		1z – 15nm	1α – 14nm	1β - 13nm			1δ - 11nm
SS	1x – 18.5nm	1y – 16nm		1z – 15nm		1α – 14nm	1β - 13nm	1γ - 12nm	1δ - 11nm	

Nodes are active half-pitch.

DRAM Capacitor Scaling Issues

IC KNOWLEDGE LLC

- DRAM capacitors are fabricated at the limits of mechanical stability (see figure on the right). A titanium nitride storage node is formed that the dielectric layer and top plate are deposited over.
- Higher K value dielectrics have lower band gaps and therefore higher leakage.
- DRAM scaling has become an optimization battle between achieving a minimum capacitance value, minimizing leakage and optimizing the peripheral circuitry.

$$C = \frac{k\varepsilon A}{t_d}$$ **Capacitance Formula**

$k = dielectric\ constant$
$\varepsilon = permitivity$
$A = area$
$t_d = dielectric\ thickness$

Band Gap Versus K

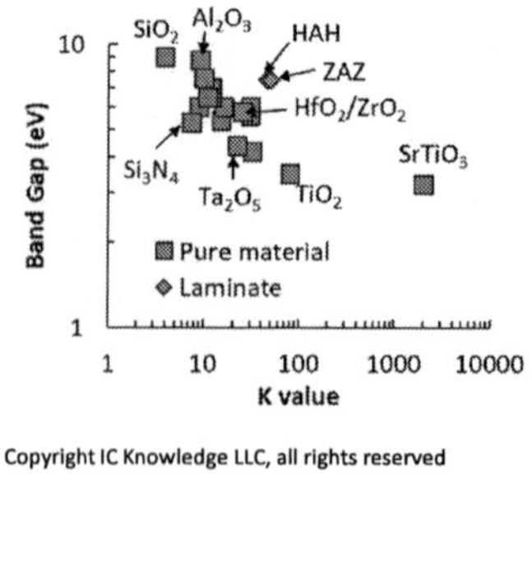

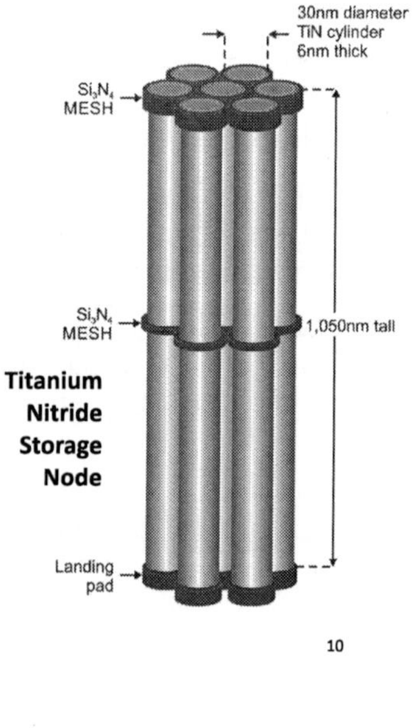

DRAM Process Challenges

IC KNOWLEDGE LLC

- Double MESH storage node support going to triple MESH – Samsung at 1z.
- More complex capacitor dielectric stacks, Micron adding Nb and HfO at 1x, Samsung adding HfO, at 1z. A breakthrough in the basic material is possible, but looks unlikely.
- EUV, Samsung at 1z, SK Hynix at 1α, Micron at 1δ.
- Air gaps for Bit Line and Interconnect Lines.
- HKMG in peripheral transistors for power and performance.
- Low-k dielectrics at 1α and cobalt caps at 1β for interconnect.
- Cylinder capacitors transitioning to pillar capacitors.
- Beyond 1δ, 3D DRAM with horizontal capacitors after 2025.
- New interconnect metals after 2025.

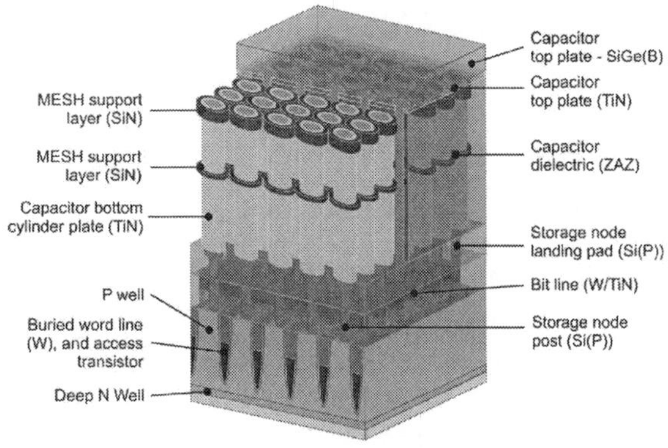

DRAM Memory Array
1x. 1y, 1z generation

DRAM Wet Process Trends

IC KNOWLEDGE LLC

- Total wet process steps by node
  - BEOL – post metal or via etch
  - BSB – backside bevel for litho or CMP
  - CMP – on-tool and post cleans
  - Etch – wet etches
  - FEOL – front end anhydrous HF, ammonium peroxide and critical cleans
  - SCCO2 – super critical CO2
  - Strip – photoresist and carbon strips
- Based on Samsung DRAM

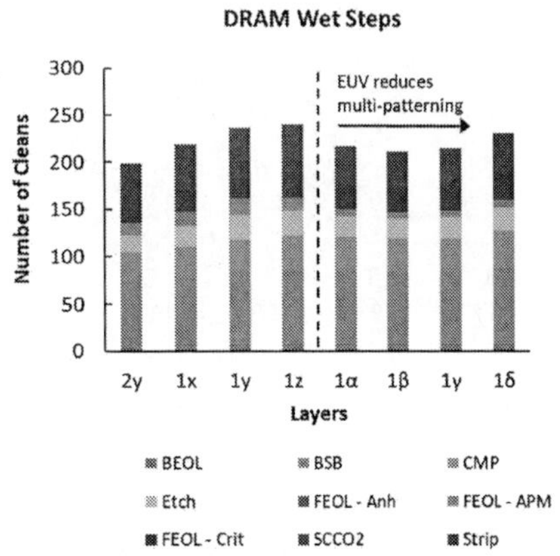

Source: IC Knowledge Strategic Cost and Price Model – 2021 – revision 00a

Logic Density Trend

IC KNOWLEDGE LLC

- Millions of transistors per millimeter squared based on a 60% NAND cells and 40% scanned flip flop cells.
- This metric applies to the logic portion of the circuit only, SRAM, analog and I/O scale differently.

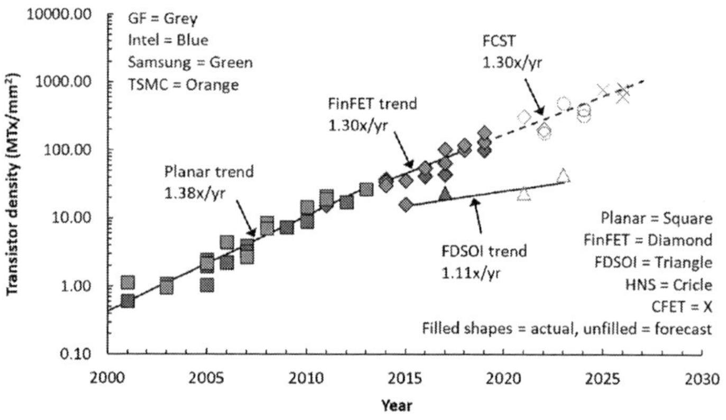

IC Knowledge and TechInsights data

Logic Nodes

IC KNOWLEDGE LLC

	2016	2017	2018	2019	2020	2021	2022	2023	2024	2025
Intel										
- Node	14	14	14	10	10SF	10SF	7	7	5	5
- Device	FF	FF	FF	FF	FF	FF	FF	FF	HNS	HNS
- TEN	13.8	13.8	13.8	7.1	7.1	7.1	4.1	4.1	2.4	2.4
- Status	Full	Full	Full	Full	Full	Full	Ramp	Full	Ramp	Full
SS										
- Node	10	10	7	5	5	5	3	3	2	2
- Device	FF	FF	FF	FF	FF	FF	HNS	HNS	HNS	HNS
- TEN	12.0	12.0	7.4	6.0	6.0	6.0	4.7	4.7	2.9	2.9
- Status	Full	Full	Full	Risk	Full	Full	Risk	Full	Risk	Full
TSMC										
- Node	10	7	7/7+	5	5	3	3	2	2	1.5
- Device	FF	FF	FF	FF	FF	FF	FF	HNS	HNS	CFET
- Status	Full	Risk	Full	Risk	Full	Risk	Full	Risk	Full	Risk

- Risk = risk starts, Ramp = production ramp, Full = full production, EN = TSMC equivalent node.
- TEN is the TSMC equivalent node.

Logic Transistors

IC KNOWLEDGE LLC

- Faucet analogy – a transistor in a logic device is like a faucet.
- A faucet turns on or off the flow of water, transistor turns on and off the flow of electric current.
- When the faucet is off you don't want it to leak, when a transistor is off you don't want it to leak.
- Leakage limits the gate length (Lg) each transistor type can achieve.
- When you turn on a facet you want a high flow of water so you can get things done, e.g. fill a bucket. When a transistor is on you want a high flow of current so you can get things done and drive a circuit at high speed.
- The effective channel width (Weff) is a key indicator of current flow capacity.

Logic Transistor Evolution

IC KNOWLEDGE LLC

- Planar to FinFET to Horizontal Nanosheet (HNS)
  - Enables shorter gate lengths = smaller cells.
  - Increasing Weff per unit are = more drive current/performance.
  - HNS can have variable sheet width across a device with narrow sheets (nanowires) for lowest leakage and wide sheets for maximum performance.

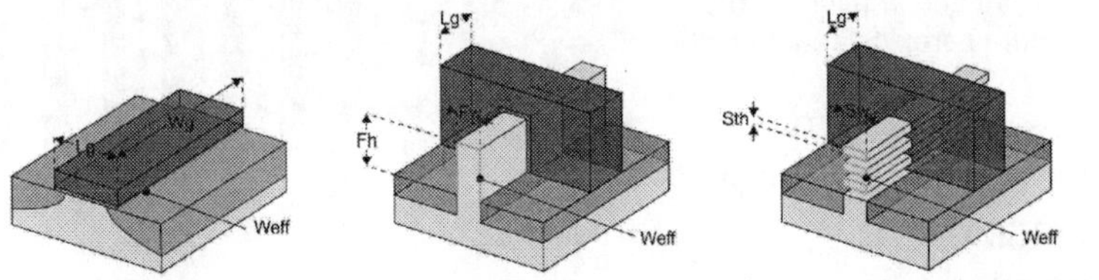

Planar	FinFET	Horizontal Nanosheet (HNS)
1 gate, Lg >= 30nm Weff = Wg	3 gates, Lg >= 16nm Weff = 2Fh + Fw Weff up to 4x planar	4 gates, Lg >= 13nm Weff = (2Sth + 2Sw) x sheets Weff up to 1.26x FinFET

Logic Process Challenges

IC KNOWLEDGE LLC

- Epitaxial growth of precisely controlled Si/SiGe layers.
- Buried power rails and backside power delivery.
- SiGe release etch without attacking Si sheets.
- Conformal dielectric and work function metal deposition.
- Creating multiple work function metals for multiple threshold voltages in the small gaps between nanosheets.
- Conformal/low resistance contacts.
- New interconnect metals and fabrication sequences such as semi damascene.

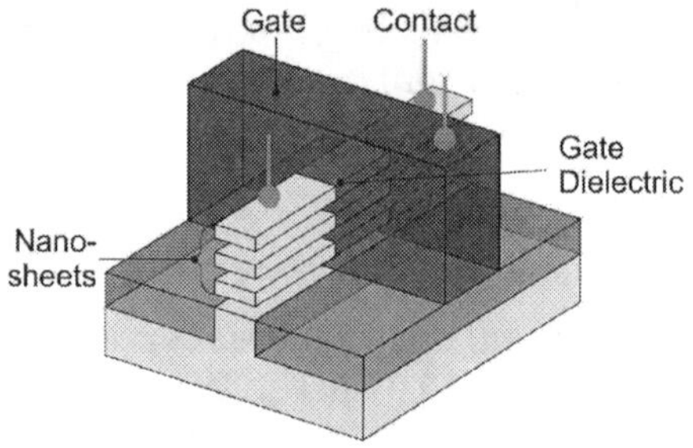

Horizontal Nanosheet (HNS)

5/27/2021 17

Logic Wet Process Trends

IC KNOWLEDGE LLC

- Total wet process steps by node
  - BEOL – post metal or via etch
  - BSB – backside bevel for litho or CMP
  - CMP – on-tool and post cleans
  - Etch – wet etches
  - FEOL – front end anhydrous HF, ammonium peroxide and critical cleans
  - SCCO2 – super critical CO2
  - Strip – photoresist and carbon strips
- Based on TSMC

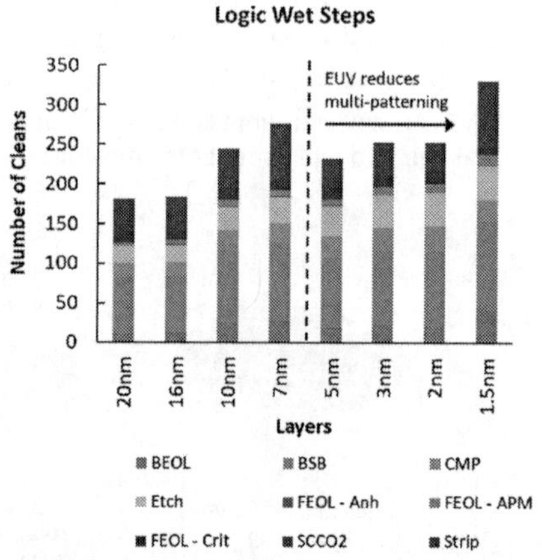

Source: IC Knowledge Strategic Cost and Price Model – 2021 – revision 00a

5/27/2021 18

Conclusion

IC KNOWLEDGE LLC

- Process scaling paths exist for all three major industry segments out to the end of the decade.
- Each new process generation tends to require more cleans.
- 3D NAND layer scaling is leading to string stacking and each additional string adds more cleans.
- The introduction of EUV replaces clean intensive multi-patterning mitigating the growth in cleans for DRAM and Logic.

5/27/2021 19

SEMES

Confidential

Controlling Static Electricity During Single Wafer Wet Cleaning Processes

Kyuhwan Chang*, Doyeon Kim, Heehwan Kim, Sunglin Lee, and Yongsun Ko

Clean Team, SEMES Co., Ltd
kyu.chang@semes.com

Contents

Confidential

- ☐What is Static Electricity?
- ☐Why is Static Electricity Important in Single Wafer Wet Cleaning?
- ☐Quantitative Analysis of Static Electricity
- ☐Controlling Static Electricity
- ☐Summary and Future Works

What is Static Electricity?

Confidential

□Static Electricity

: An **imbalance** of electric charges within or on the surface of a material.

: The charge remains until it is capable to move away by means of "**electric current**" or "**electrical discharge**".

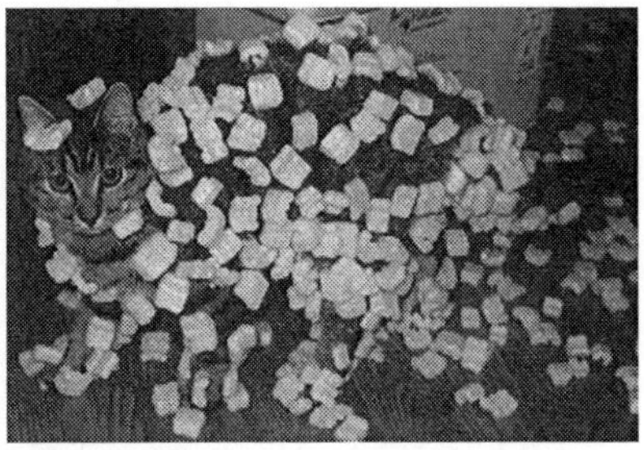

Common Examples of Static Electricity in Elementary School

① Balloon (or cat) → ② Rubbing on a surface → ③ Static Cling!

Materials + **Movement** → **Cling!**

Where Does Static Electricity Come From

Confidential

□Single Wafer Cleaning Processes

1. **Materials** – Fluoropolymer-based materials (non-conductor), Si wafer
2. **Movements** – Spinning, Spraying, Dispensing chemicals and deionized water, Swinging nozzles, Lifting up and down bowls

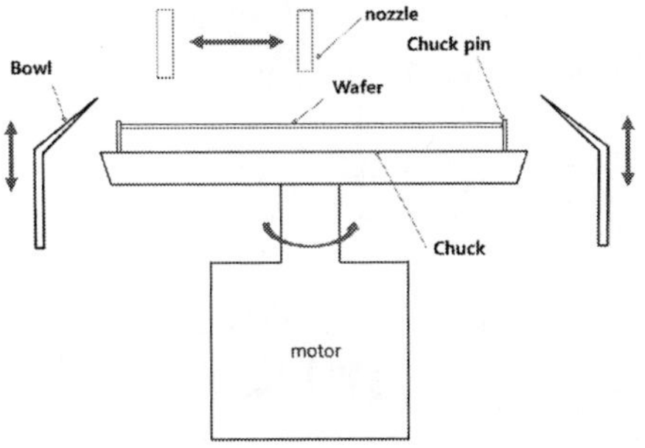

A Schematic Diagram of a Single Wafer Cleaning Chamber

A Si wafer could be in danger due to Static Electricity!

Why is Static Electricity Important

Confidential

□Indeed, wafers are in danger because of static electricity in Single Wafer Cleaning tool!

3D Surface Voltage Maps of Two Different Cleaning Methods

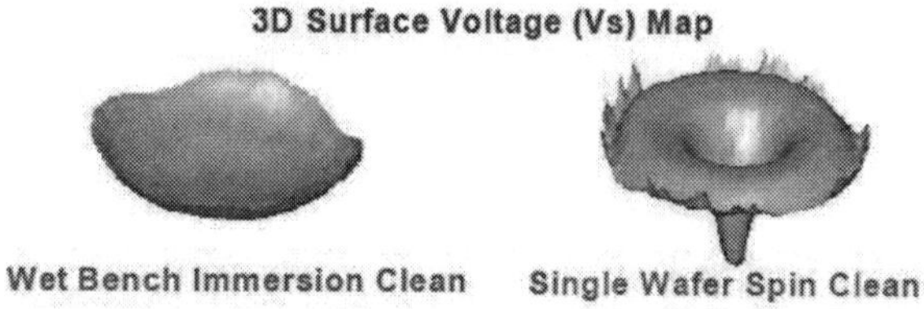

Charge to Breakdown Maps of the Gate Oxides

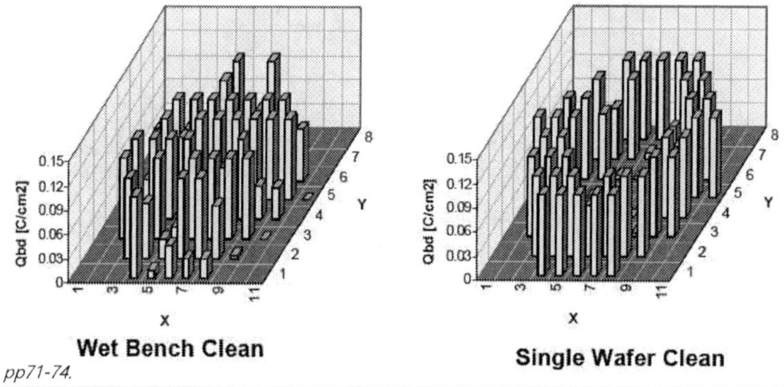

Lin et. al., Solid State Phenomena, Vol. 187 (2012) pp71-74.

Why is Static Electricity Important

Confidential

□Electrostatic Discharge (ESD) CAN cause fatal defects such as "Hot Spot" or "Pitting" on wafers.

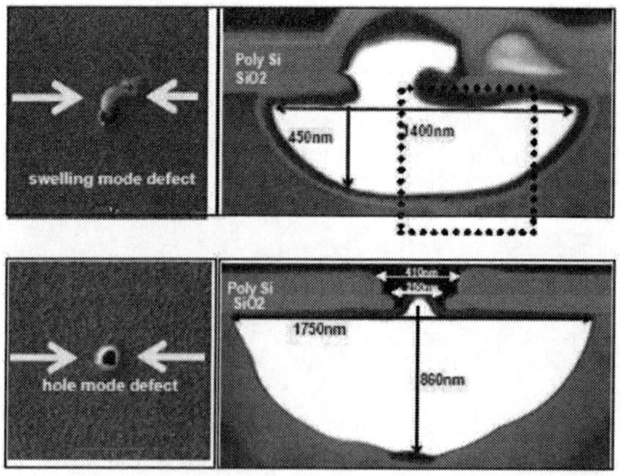

Poly-Si
Thermal oxide (or Nitride)
Sub-Si

Poly-Si
Thermal oxide (or Nitride)
Sub-Si

Hagimoto et. al., Solid State Phenomena, Vol. 145-146 (2009) pp185-188.

Why is Static Electricity Important

Confidential

☐Single Wafer Cleaning Processes

1. **Materials** – Fluoropolymer-based materials (non-conductor), Si wafer
2. **Movement** – Spinning, Spraying, Dispensing chemicals and deionized water,

   Swinging nozzles, Lifting up and down bowls
3. **Static Electricity** Cling! Or Discharge

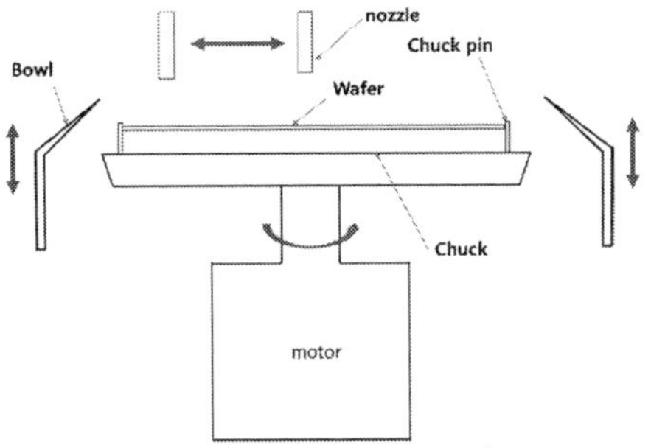

A Schematic Diagram of a Single Wafer Cleaning Chamber

The "Perfect Playground" for Static Electricity

Where Does Static Electricity Come From

Confidential

☐The Triboelectric Effect

Certain materials are electrically charged when they are separated with which they were in contact.

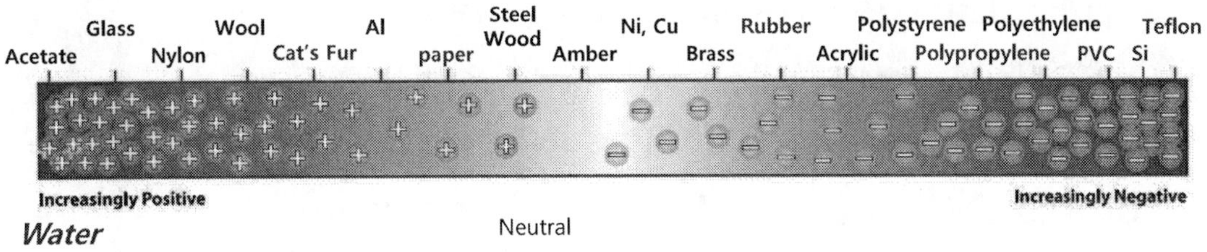

- Most **Fluoropolymer materials are negatively charged** in the series.
- **Water is top of the list on the positive** side of the series.

Zou H. et. al., "Quantifying the triboelectric Series", Nature Communications, 10(1): p1427 (2019).

Burgo T.A.L., et. al., "Where is Water in the Triboelectric Series", Journal of Electrostatics, Vol. 80, pp 30-33 (2016)

Where Does Static Electricity Come From

Confidential

□Triboelectric Effect during Single Wafer Cleaning

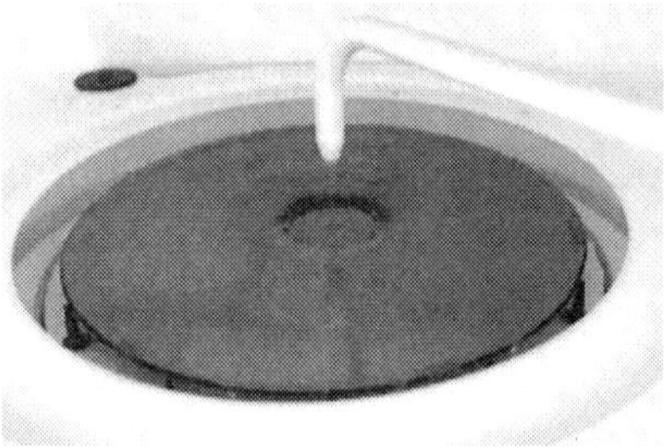

DIW (or Chemical) with Pipings

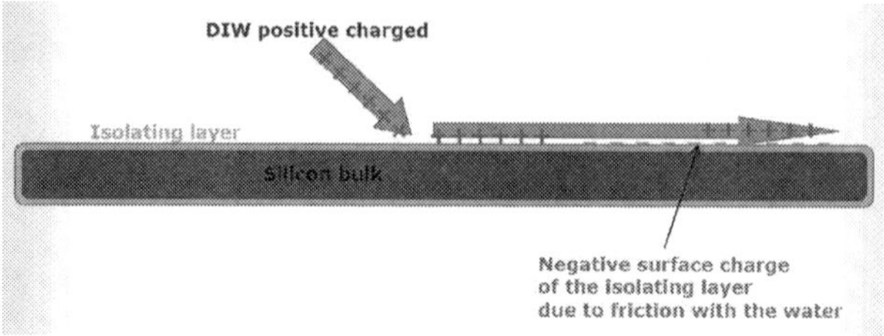

DIW (or Chemical) with Wafer (or Chuck Surface)

DIW (or Chemical) with Bowls after Processing

SEMES

Quantitative Analysis of Static Electricity

Confidential

□Liquid Charge _Coulomb (Faraday cup)

- A cup type measurement for the amount of charge in the solution.
- Dependent on the amount and speed of the liquid dispensed.

□Surface Charge _Voltage (Field meter)

- Used for "non-contact" measurement of the electrostatic field.
- Fitted form measuring the surface of insulator materials.

□Voltage Detection _Voltage (voltage meter)

- Monitors electrostatic field voltages in multiple locations.
- Measures the voltage changes in real time.

□E-Wafer Measurement*

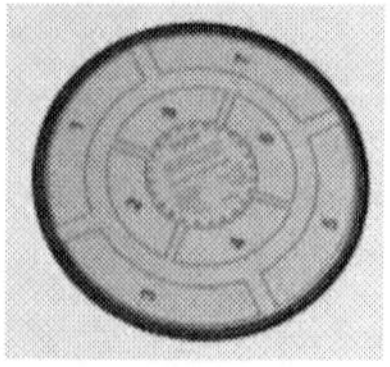

* *http://www.estion-tech.com/products/e-wafer*

SEMES

Quantitative Analysis of Static Electricity

Confidential

□DIW (or Chemical) with Pipings

- Measurement of discharge charge using Faraday cup
- Evaluation condition: PFA tube, DIW Dispense

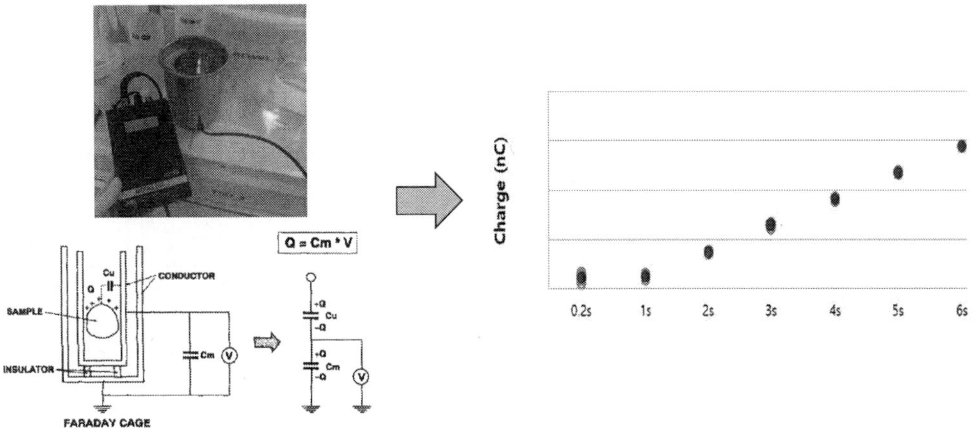

Measurement of Static Charge of DI Water by Faraday Cup (or Faraday Cage)

Ohmi T. et. al., IEEE Trans. On Semicon. Manufac., Vol 7, No. 4, pp440-446 (1994)

Quantitative Analysis of Static Electricity

Confidential

□DIW (or Chemical) with Pipings

- Effects of Wafer RPM and CO_2W
- Evaluation condition: PFA tube, DIW (or CO_2W) Dispense

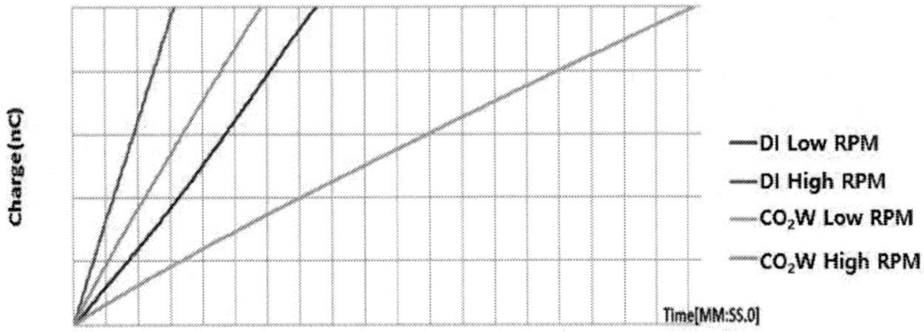

- The amount of charges linearly increases with the dispense time.
- The slope linearly related to the RPM during the dispense step.
- Use of CO_2W reduces the amount of charge in dispensed DI.

Quantitative Analysis of Static Electricity

Confidential

□DIW (or Chemical) with Pipings

- Effect of the pipe material on the electrostatic charge of DIW
- Measure the amount of charge to pipe materials and GND

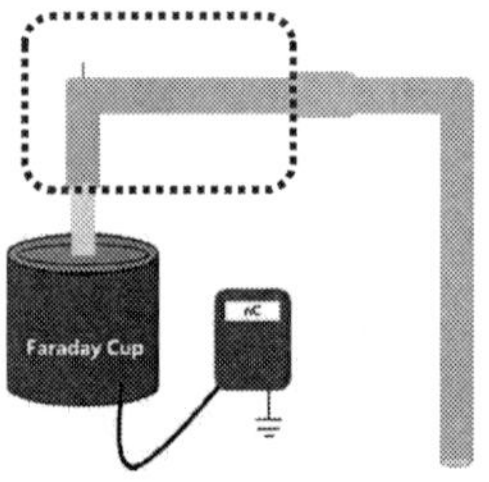

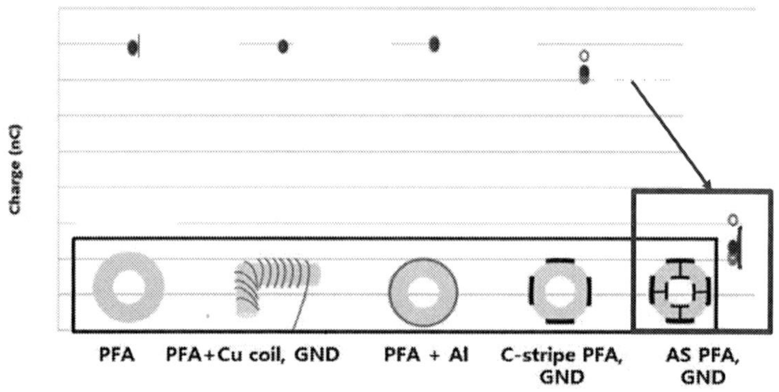

Faraday Cup Measurement Results of DI Water Dispensed through Different Type of Pipes

Quantitative Analysis of Static Electricity

Confidential

□DIW (or Chemical) with Pipings

- Measure the effect of different resistivities and GND on the static charge.

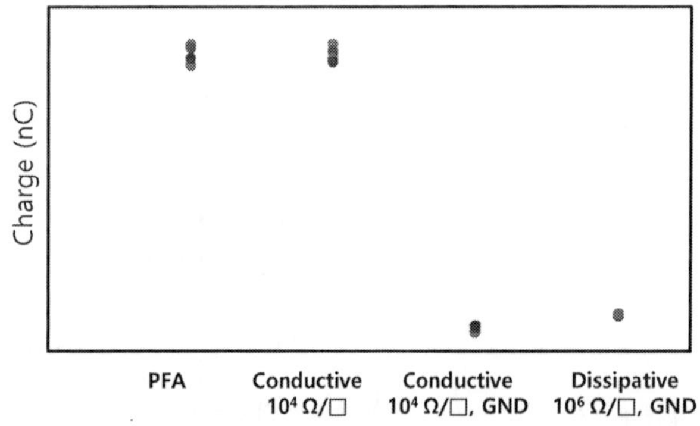

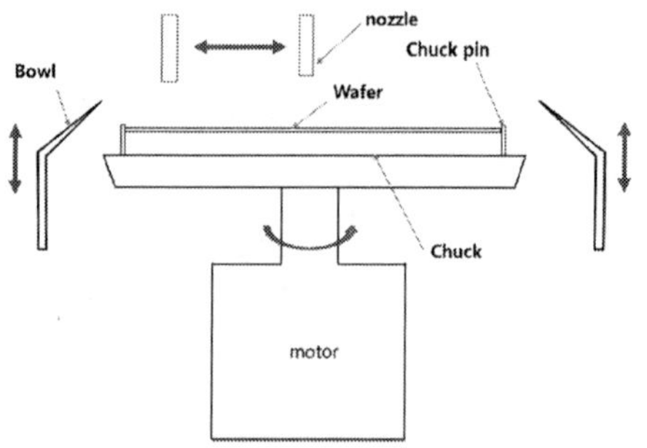

- Grounding is required to discharge static electricity charged with solution.
- The material in the dissipative range can be used to prevent reverse charging.

Quantitative Analysis of Static Electricity

Confidential

□Single Wafer Cleaning Processes

1. **Materials** – Fluoropolymer-based materials (non-conductor), Si wafer
2. **Movement** – Spinning, Spraying, Dispensing chemicals and deionized water, Swinging nozzles, Lifting up and down bowls
3. **Static Electricity** Cling! Or Discharge

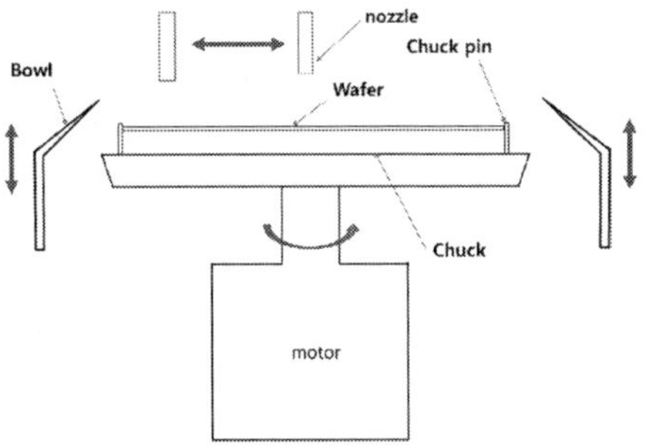

A Schematic Diagram of a Single Wafer Cleaning Chamber

SEMES

Quantitative Analysis of Static Electricity

Confidential

□Field meter results of chuck surface after processes

- Different solutions dispensed on the chuck surface
- Static Electricity (Volt) measured by a Field Meter

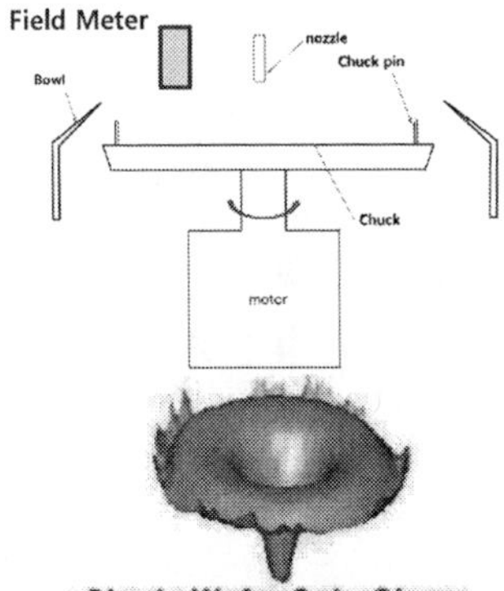

Single Wafer Spin Clean

Chemical	Resistance (ohm*cm)	Relative Scale of Static Electricity			
		Center	~	EDGE	Δ
DIW	18.2 M	-0.92	~	+0.06	0.98
SC1	≒ 0.03	-0.48	~	+0.03	0.51
Dilute HF	≒ 2.0	+0.1	~	0	0.1
CO_2W	~0.1 M	-0.004	~	0	0.004

- The same tendency of negative charge is observed by a field meter after a DI process at the center of a chuck made of a Fluoropolymer material.

SEMES

Quantitative Analysis of Static Electricity

Confidential

□Field meter measurement results of wafer surface

- Static Electricity (Volt) of a wafer monitored by a Field Meter while running a process recipe
- Static Electricity affected by all the movements such as nozzle movements, wafer rpm, flow rate of each step

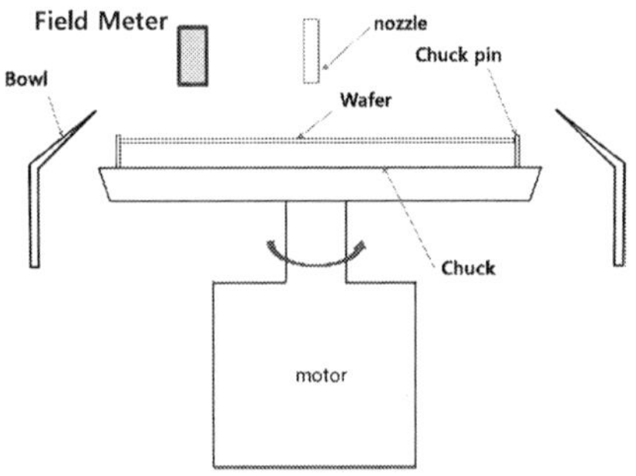

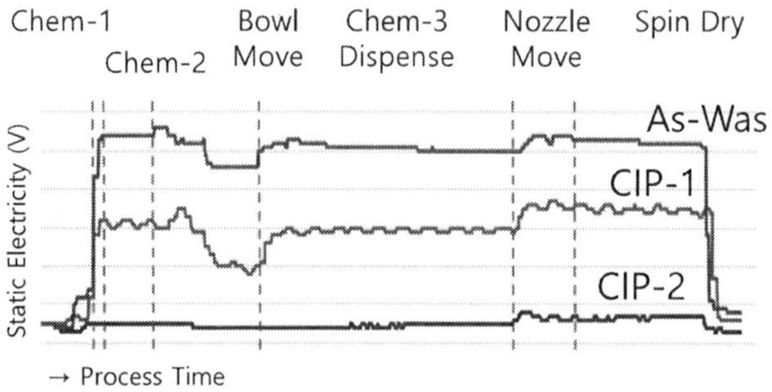

SEMES

Quantitative Analysis of Static Electricity

Confidential

□Field meter measurement results of wafer surface

- Static Electricity affected by all the movements such as nozzle movements, wafer rpm, flow rate of each step
- Static Electricity (Volts) of a wafer monitored by a Field Meter while running a process recipe

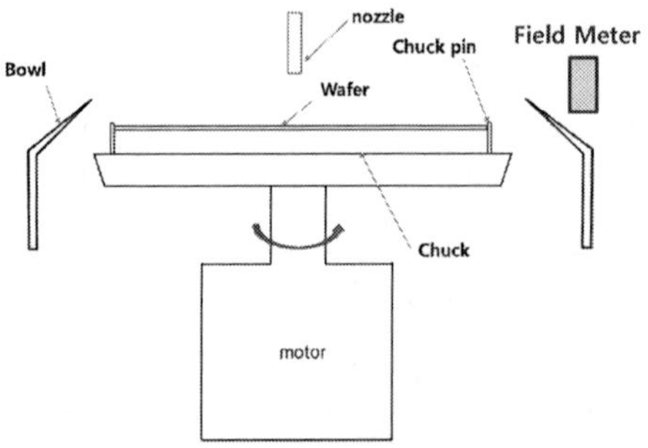

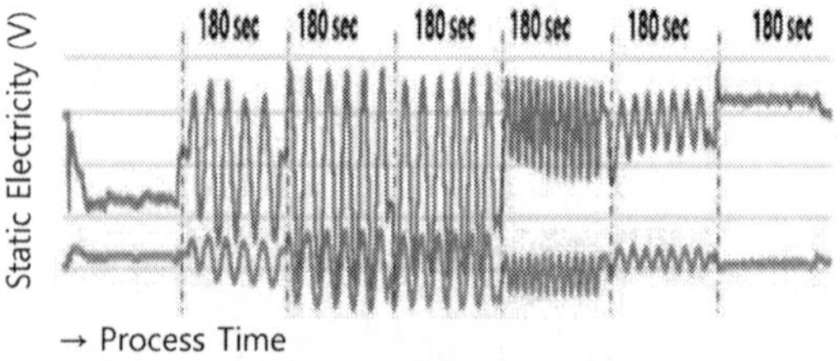

- A "waveform" behavior of static electricity observed from the outside of bowl during the process.

SEMES

Quantitative Analysis of Static Electricity

Confidential

□Toroidal Magnetic Field Formed Around the Chuck

- The "sine-wave" behavior only seen on a spinning chuck with DI being dispensed.
- Toroidal magnetic field confirmed with compass observation
- Numerical modeling

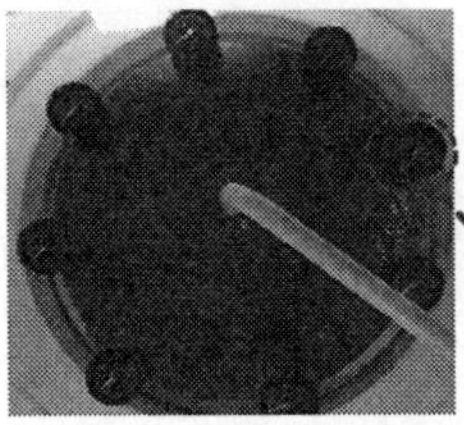

Setup to Visualize Toroidal Magnetic Field

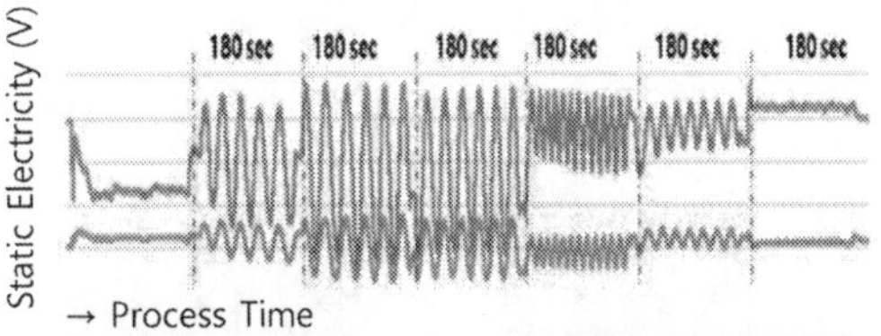

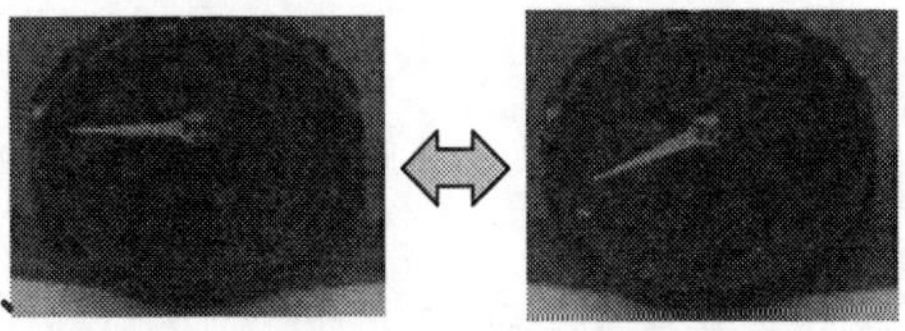

Same cycle of compass and Sine wave

Quantitative Analysis of Static Electricity

Confidential

□Single Wafer Cleaning Processes

1. **Materials** – Fluoropolymer-based materials (non-conductor), Si wafer
2. **Movement** – Spinning, Spraying, Dispensing chemicals and deionized water, Swinging nozzles, Lifting up and down bowls
3. **Static Electricity** Cling! Or Discharge – ESA or ESD

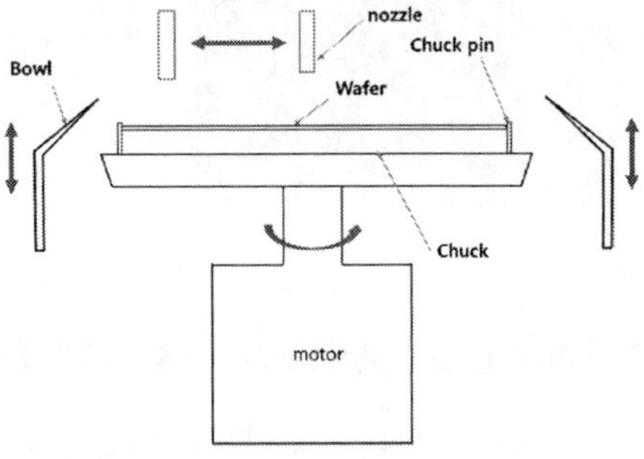

A Schematic Diagram of a Single Wafer Cleaning Chamber

Electro-Static Attraction (ESA) - Particles

Confidential

☐ **The correlation between wafer static electricity and the particles around the wafer**

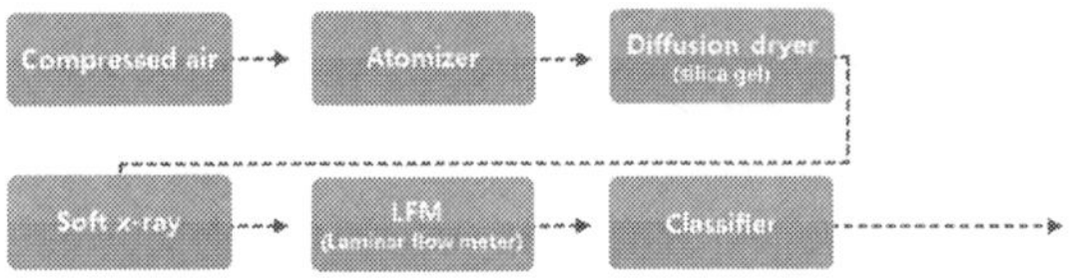

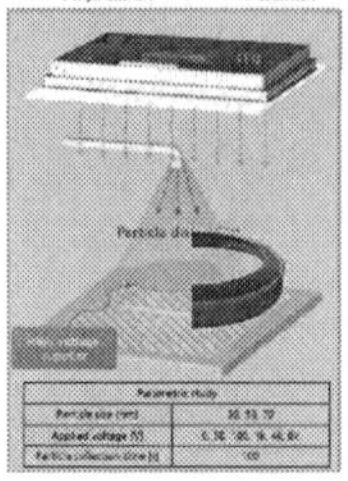

Evaluation of particle adsorption according to voltage applied amount

Particle Deposition Velocity (v_d)

$$V_d = \frac{J}{C_a} = \frac{N_{de}}{C_a A_{de} t}$$

V_d : Particle deposition rate

N_{de} : Number of deposited particles

C_a : Concentration of surface particles

A_{de} : Cross section / t : Exposure time

PARTICLE ATTRACION EFFICIENCY

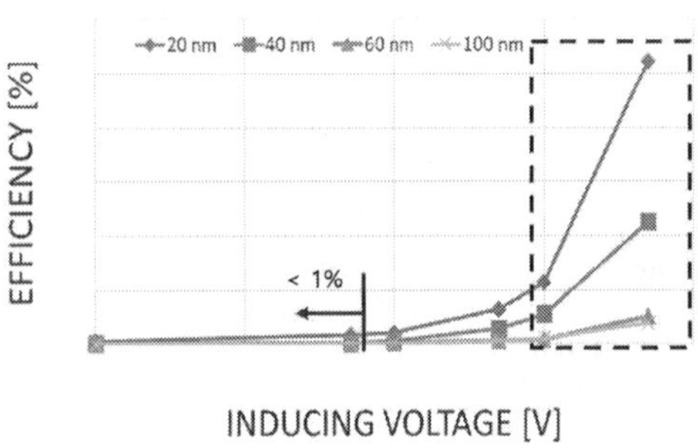

PARTICLE DEPOSITION VELOCITY

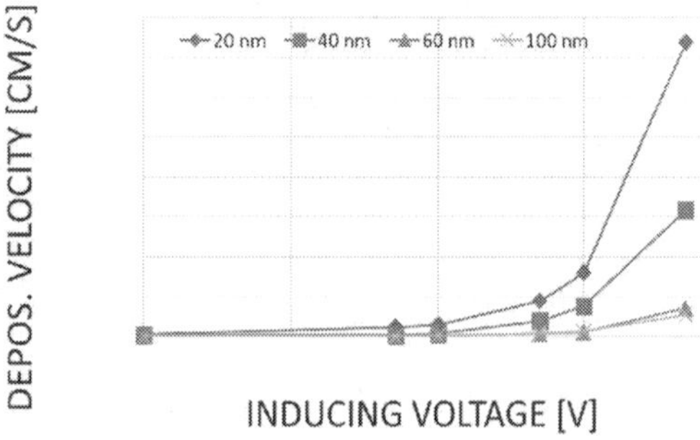

Electro-Static Discharge (ESD) - Defects

Confidential

☐ **Potential differences between two different spots on a wafer often causes this type of defects.**

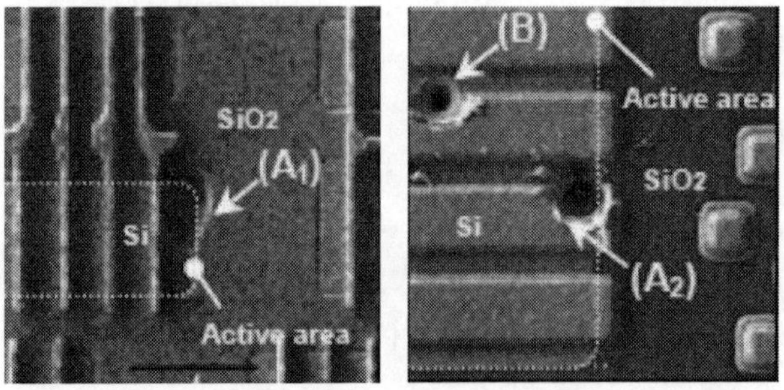

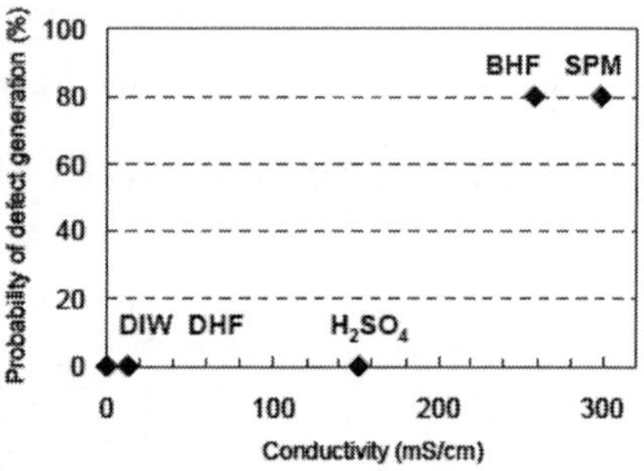

☐ **The number of defects increases for the highly conductive solutions such as BHF and SPM.**

Hagimoto et. al., Solid State Phenomena, Vol. 145-146 (2009) pp185-188.

Controlling Static Electricity

Confidential

□Grounding

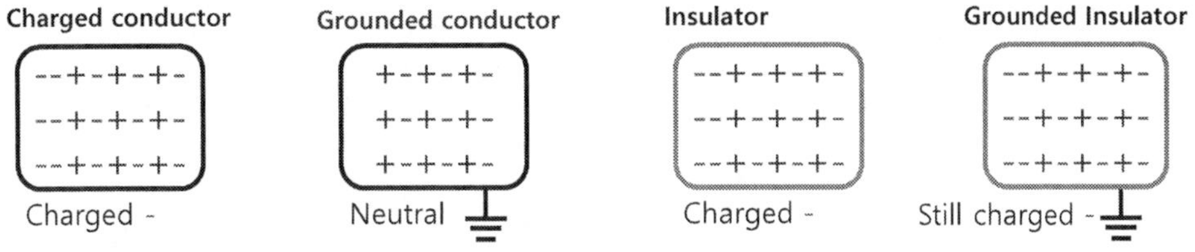

□Dissipative (Anti-Static) Materials

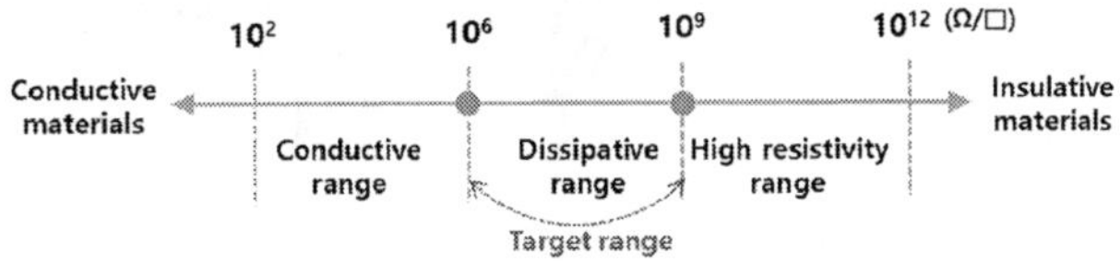

□Ionizer

[Corona Ionizer]

[DC ionizer]

[X-ray ionizer]

SEMES

Controlling Static Electricity

Confidential

□Toroidal field disappeared with a dissipative bowl

- Bowl is replaced with another made of dissipative material
- The surface resisitivity is in the range of $10^{6\sim9}$ Ω/□.
- Effective even WITHOUT ground.

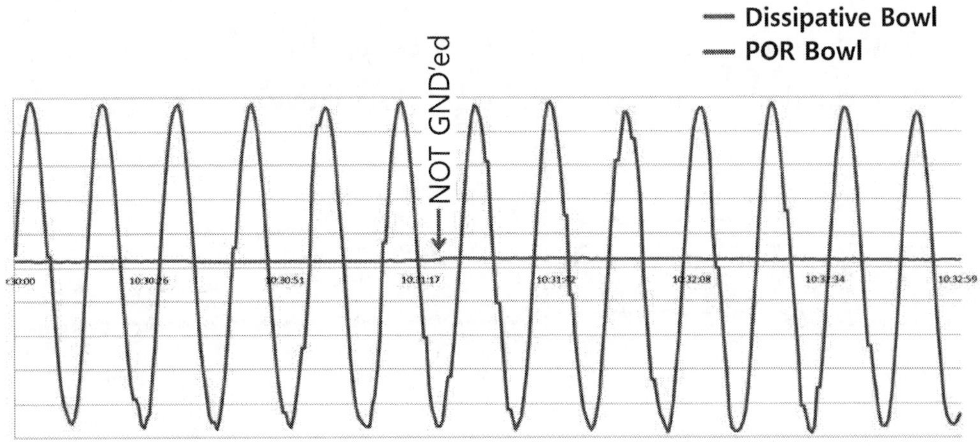

SEMES

Controlling Static Electricity

Confidential

□Spin chuck with a surface made of a dissipative material

- No static potential measured with field meter when a spin chuck is made of a dissipative material.

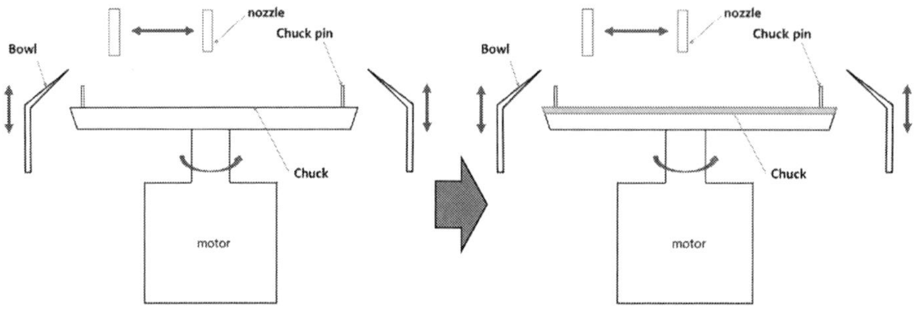

	Relative Scale of Static Electricity		
Chuck materials	Center	Middle	Edge
POR chuck + DIW	- 1.32	- 1.16	- 0.72
Dissipative 1+ DIW	-0.02	-0.004	-0.02
Dissipative 2+ DIW	-0.04	-0.02	-0.02

SEMES

Controlling Static Electricity

Confidential

□Spin chuck with a surface made of a dissipative material

- A test recipe ran with a wafer with a field meter monitoring the wafer surface.
- Compared to POR chuck,
  ~ 30% reduction of surface voltage by Dissipative 1.
  ~ 88% reduction of surface voltage by Dissipative 2.

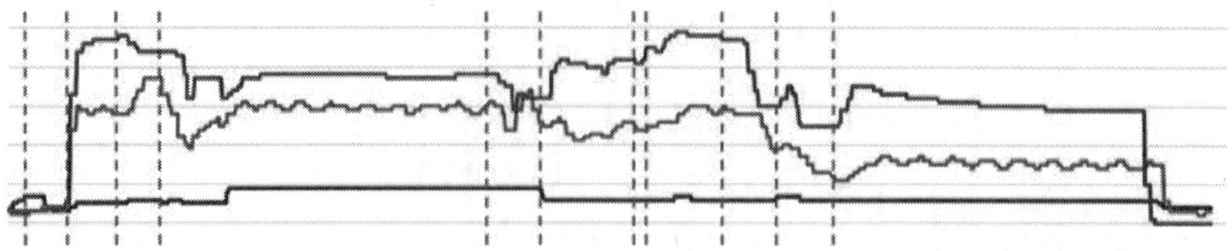

POR Chuck
Dissipative 1
Dissipative 2

SEMES

Summary and Future Works

Confidential

- ☐ **A single wafer wet cleaning → "Perfect Playground for Static Electricity" due to ;**
  - (1) materials used in terms of triboelectric series
  - (2) dispensing DI and chemicals, spinning a chuck, moving bowls and nozzles
  - (3) causes ESA and/or ESD if NOT properly controlled.
- ☐ **The tolerance of device for static electricity seems smaller and smaller with the design rule.**
- ☐ **Static Electricity MUST be controlled in a proper level, if not eliminated, by process optimization, material selection, ionization, and proper connection to grounding.**
- ☐ **Further study needs to be done on the effect on chemical reactions.**

SEMES

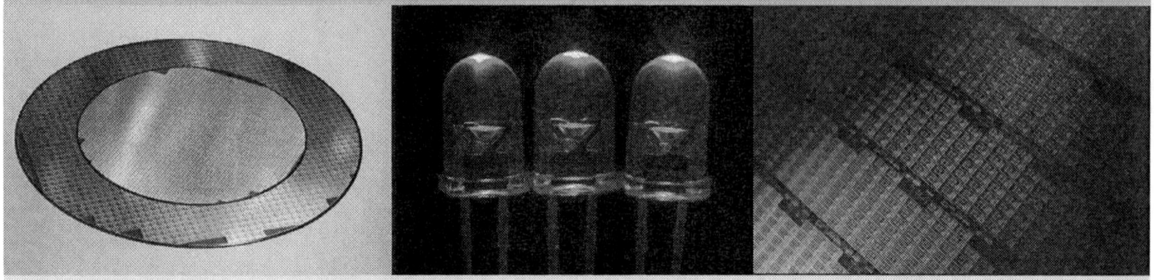

Cleaning Materials Market

Business of Cleans 2021
June 8, 2021

David Maloney

www.linx-consulting.com
617.273.8837• 973.698.2331

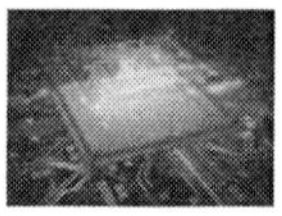

Outline of Presentation

- Semi Demand
- Macroeconomic environment
- Geopolitical environment
- Boats & planes
- Cleans – selected business & technical challenges
- Selected issues for discussion

See Beyond the Horizon

www.linx-consulting.com

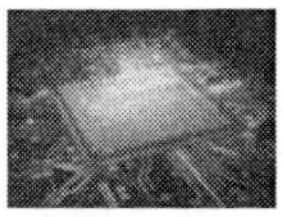

Semi Demand

- Consumer demand for products incorporating semiconductor devices is high
  - COVID19
    - Phones, laptops, displays, routers, webcams....
    - "...cycle of explosive growth in semiconductors."
  - Part of ongoing proliferation of semi-intensive applications
    - Automotive
    - White goods
  - Outlook uncertainty, stockpiling, whiplash
  - Disruptions to supply
  - BCP?
- Which creates some notable dynamics
  - The prices of new and used cars
  - 200 mm
  - Pricing and supply challenges in cleans

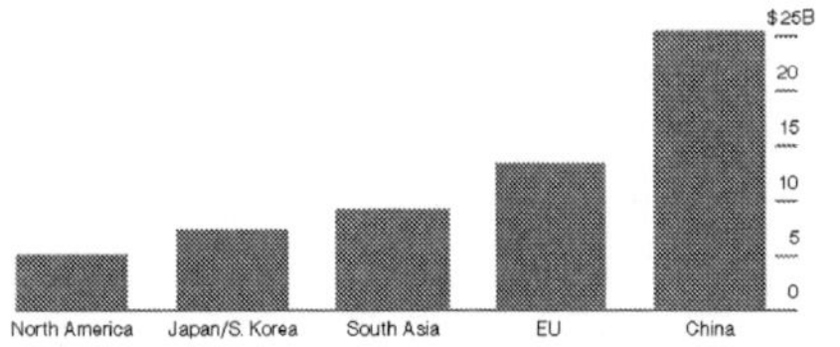

Source: AlixPartners
Note: Estimates based on current rate of volume losses

Automakers 2021 Losses – est. $110B as a result of chip shortages

See Beyond the Horizon

www.linx-consulting.com

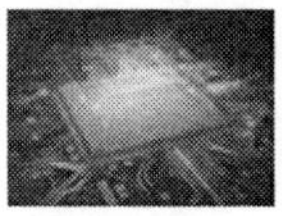

Semi Demand - Technology Drivers

- 5G
  - Mobile device production
  - Network infrastructure build out
- Server infrastructure for AI, Machine Learning, VR/AR and IoT
  - Processor heavy
  - High Memory demand
- Automotive
  - Increase in automotive electronics
  - ADAS
  - Autonomous vehicles
- IoT
  - Distributed sensors
  - Edge cloud
  - Cloud computing
- AR / VR
  - Novel environments for improved performance

- Strong growth of processors and processing power to support continuing digitalization - ADAS, AI, IoT, and Quantum computing.
- Mobile devices exceed memory and processor speeds for legacy PCs to support 5G.
- 5G build out requires infrastructure investment to support a dense network.
- Sensors, cameras and local processing supports digitalization.

www.linx-consulting.com

The 200 mm Environment

- The supply of 200 mm manufacturing capacity is maxed-out
  - Automotive
  - COVID19
- Capacity is being added
- Not many 200 mm equipment makers
  - High demand + limited supply = …
  - Robust refurb market
- And where will all the 200 mm wafers come from?
- Option to migrate to 300 mm (although 200 mm is operationally inexpensive)
  - 180 nm → 130 nm
  - Al → Cu interconnect

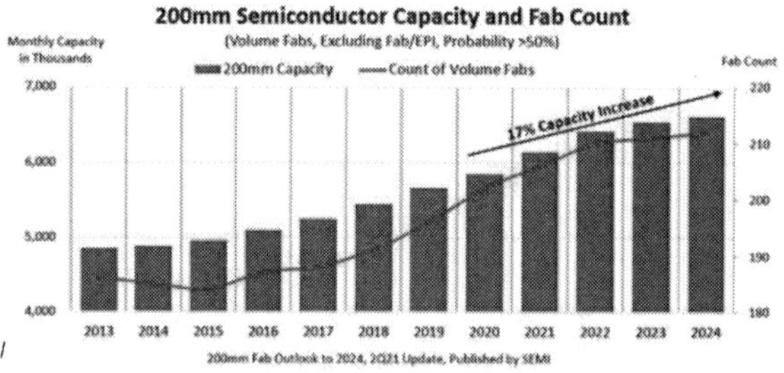

SEMI

See Beyond the Horizon

www.linx-consulting.com

Macro and Geopolitical Environment

- Supply Chain Resilience, risk mitigation and (re)domestication of semi manufacturing
  - US
    - Intel AZ
    - TSMC AZ
    - CHIPS Act and executive order
  - EU
    - Discussions with Intel
    - “Digital Compass”; “Leading the Digital Decade” 06/2021
  - China
    - Ownership in fab companies
    - Semi a longstanding priority; series of tax changes to spur semi growth
  - Korea
    - “K-Belt Semiconductor Strategy” 5/12/2021
  - Japan, India…..
    - Japanese Ministry of Economy, Trade and Industry report 6/4/2021
- Trade Protectionism – what do the next few years look like?
- What is the weather forecast?
  - Electricity
  - Water

See Beyond the Horizon

www.linx-consulting.com

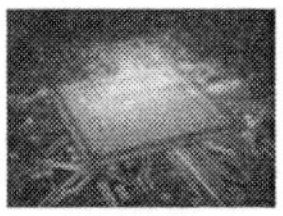

H.R.7178 - CHIPS for America Act (6/11/2020)

- This bill establishes investments and incentives to support U.S. semiconductor manufacturing, research and development, and supply chain security.
- Specifically, the bill provides an income tax credit for semiconductor equipment or manufacturing facility investment through 2026. The bill also establishes a trust fund to be allocated upon reaching an agreement with foreign government partners to promote (1) consistency in policies related to microelectronics, (2) transparency in microelectronic supply chains, and (3) alignment in policies towards nonmarket economies.

Executive Order: "Securing America's Critical Supply Chains" (2/24/2021)

- Section 3 b (i) : The Secretary of Commerce, in consultation with the heads of appropriate agencies, shall submit a report identifying risks in the semiconductor manufacturing and advanced packaging supply chains and policy recommendations to address these risks. The report shall include the items described in section 4(c) of this order.

See Beyond the Horizon

www.linx-consulting.com

Boats & Planes

Boats

- "space is full until mid-June and rates are still escalating with premium surcharges that continue to increase nearly every other day and without notice"
- "Ongoing severe congestion at Hutchison's Yantian International Container Terminal (YICT) in Shenzhen has prompted the port to extend its temporary closure to export containers"
- Exacerbated by COVID19 & Suez Canal

Planes

- Shippers are shifting greater volumes of US inbound airborne shipments to secondary airports to avoid ground handling bottlenecks at main air cargo gateways, which are being overwhelmed by surging demand.
- US airport hubs such as Los Angeles and Chicago are being inundated with import volume as a resurgent economy and high levels of consumer spending push up demand. Meanwhile, smaller, secondary airports are building their list of regularly scheduled flights from new airlines.
- Shifting volumes to these alternate regional airports, such as Rickenbacker International Airport (LCK) near Columbus, Ohio and Dallas-Fort Worth, is more efficient for shippers

See Beyond the Horizon

www.linx-consulting.com

Boats & Planes

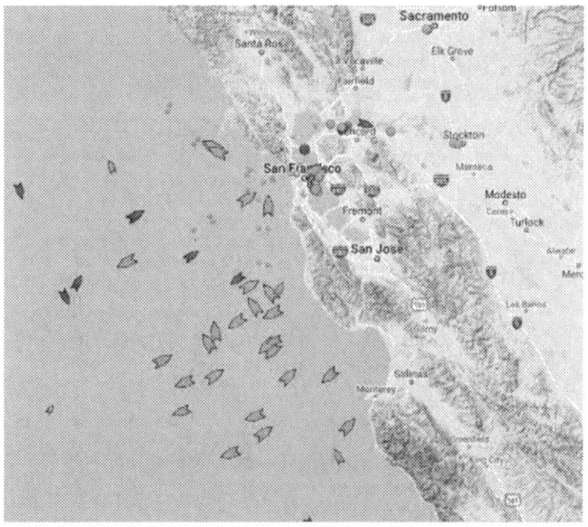

6/4/2021

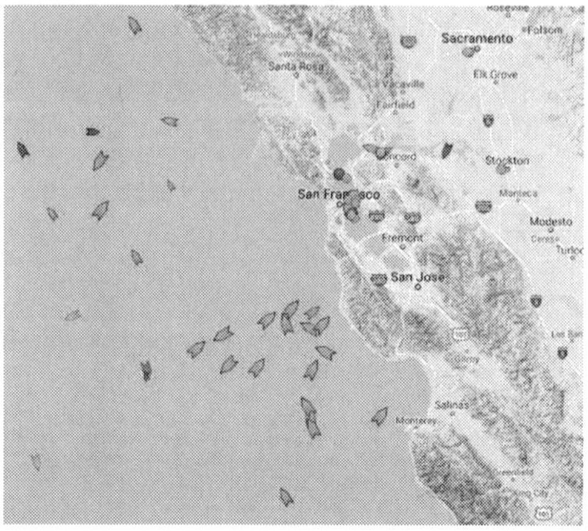

6/7/2021

Major Cleaning Challenges

Business-related:

- Very tight for many popular solvents, acids and bases routinely employed in semi cleans as well as other consumables that are critical inputs
- Due to semi demand, competing demand in other markets and the weather
- Supply chain constrictions – not all of which are COVID-dependent, related to shipping and other constraints

Major Cleaning Challenges

- Novel architectures and associated new materials
  - For example:

Device Architecture Roadmap

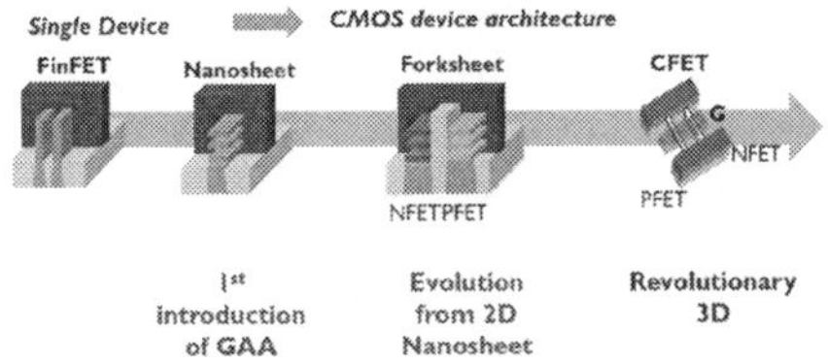

5T cells in the future

5T With Buried Power Rail (BPR)

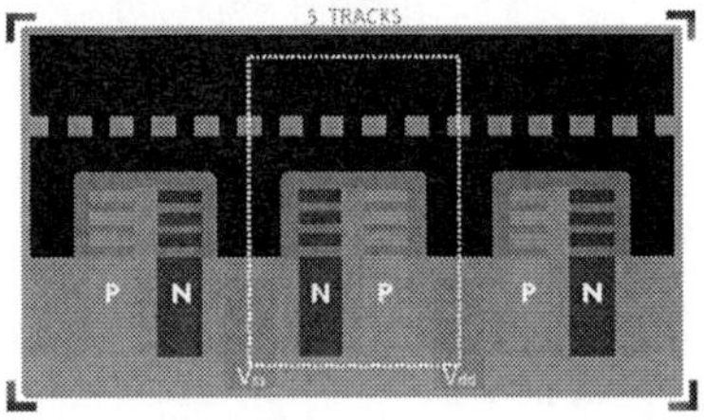

EVOLUTION FROM FINFET TO NANOSHEET

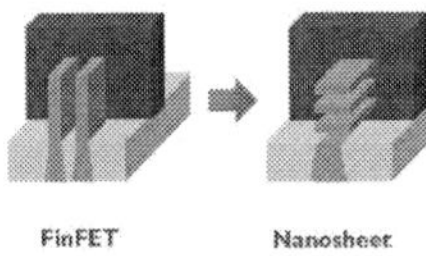

Nanosheet Advantages

- Maximized effective width → Drive current increase
- Improved short-channel control due to gate-all-around
- Variable device width for design flexibility

What's new

- **Single fin** architecture w/ stacked **nanosheet channel**
- Scaling of NP spacing by **Forksheet** or **CFET** (not shown).
- High aspect ratio **buried power rails** in STI/substrate

Source: SPCC 2020 / IMEC

See Beyond the Horizon

www.linx-consulting.com

Major Cleaning Challenges

- Particles
  - Features are ever-smaller (x-y) and deeper (z). How to remove smaller & smaller particles from deeper & deeper holes?
  - One approach – do not allow particles into those places → filtration
  - So what is a particle at modern semi scale?
    - Advanced metrology can detect FM at ~ 10 nm
    - Oligomers, amorphous SiO_2....fall into this category
  - Filtration is not merely about "particles" – advanced functionalization, control of surface affinity (charge, Van der Waals forces) and other innovations required to meet challenges

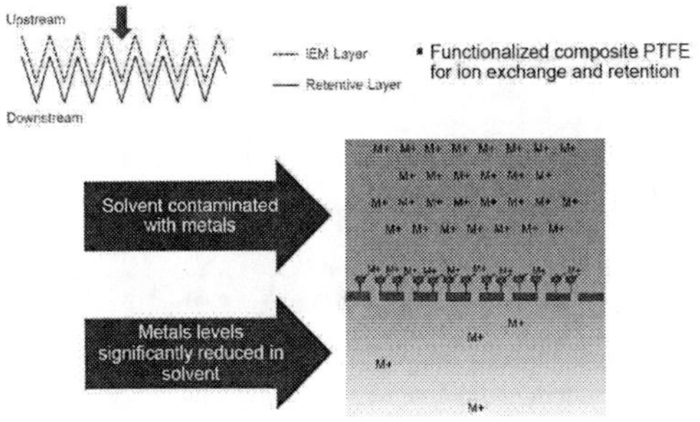

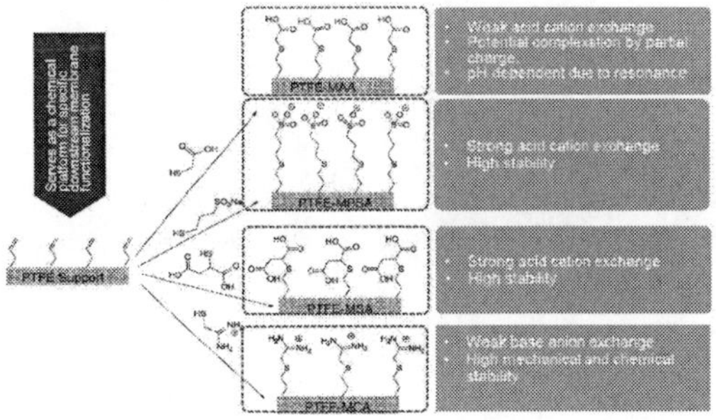

Source: SPCC 2020 / Pall Filter

See Beyond the Horizon

www.linx-consulting.com

Major Cleaning Challenges

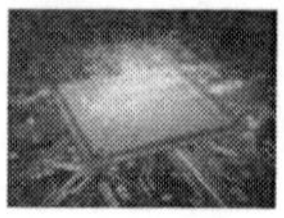

- Drying
  - Features are ever-smaller (x-y) and deeper (z). How to dry effectively?

IPA Drying
Current & Looming Issues
- Purity – monomolecular organic contamination
- Rigid supply chain
- Not sufficiently extendable

Polymer Drying
- Spin-on polymer after cleaning
- This polymer is thermally-decomposable
- Decomposition products are volatile and are sublimated away

Surface Modification Drying
- A combination of surface silylation chemistry with a series of organic solvents
- Optionally followed by a surface recovery step

Sublimation Drying
- Small-molecule organic coated onto surface after cleaning *or*
- Solid formed in gas-phase reaction that fills trenches
- Solid is volatile and is sublimated

CO_2 Drying

Major Cleaning Challenges

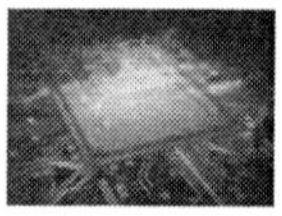

- Wet Etch
  - Legacy – e.g., Si_3N_4 Wet Etch
    - Long, hot, batch, tradeoff between e/r and selectivity to SiO_2

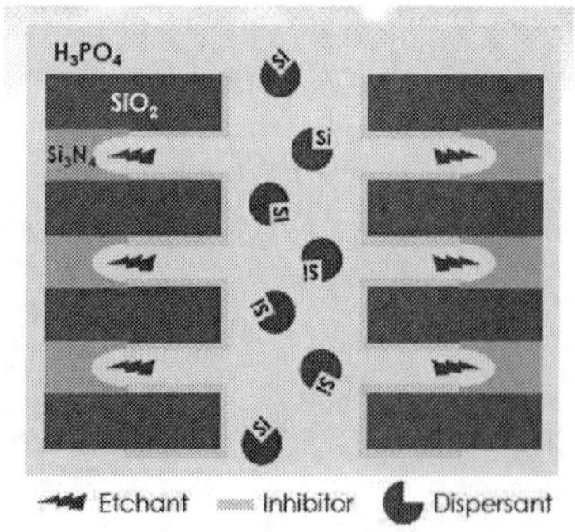

Source: SPCC 2020 / Entegris

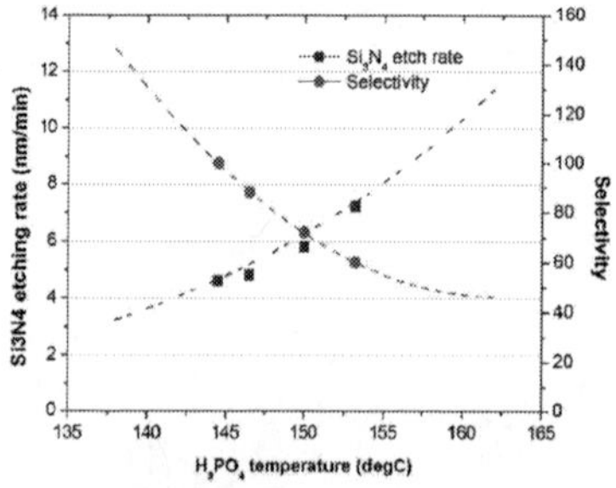

Figure 7. The Si_3N_4 etching rate and Si_3N_4/SiO_2 etching selectivity against the H_3PO_4 temperature.

Source: J. Electrochem. Soc., ***165*** *(4) H3187-H3191* (2018)

Major Cleaning Challenges

- Wet Etch
  - TiN selective etch; SiGe selective etch; *(etc.)*
  - Packaging-level selective metal etchants

Source: SPCC 2020 / Avantor

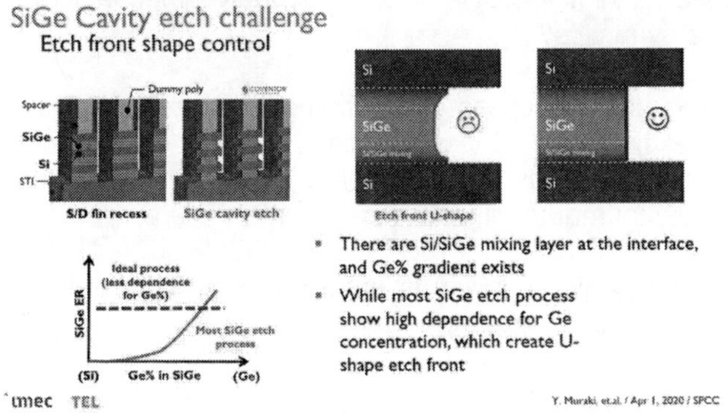

See Beyond the Horizon

www.linx-consulting.com

Outlook

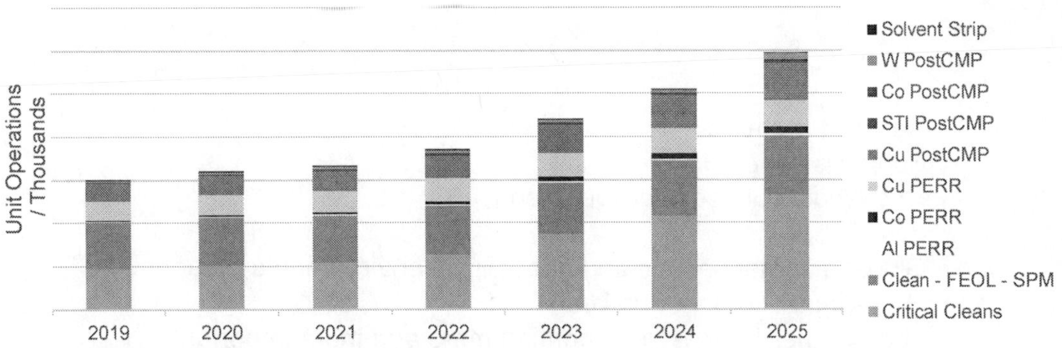

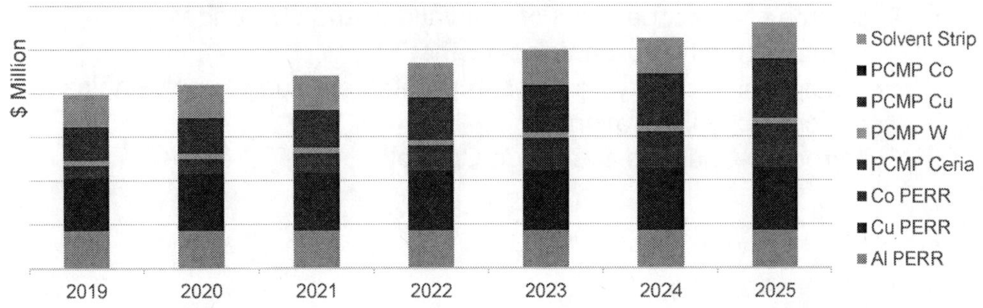

See Beyond the Horizon

www.linx-consulting.com

Business of Cleans – Selected Issues for Discussion

1. Release control, specifications for materials, chemicals, blends: in a world of ship-to-control, and in the face of a wide range of methodologies for managing release of product, is there room for some standardization? Are there additional mechanisms to be put in place to the overall industry's benefit?

2. There exists a historic (and growing) challenge of managing materials from upstream suppliers and markets who are not primarily in semiconductor-related businesses, and sometimes don't need or want to be – what collaborative mechanisms can we put in place to manage these dynamics?

3. Change control systems, the processes and documentation required to effect change in products supplied to leading fabs vary widely among them. Understanding how critical these processes are, is there a way to impart some level of uniformity to them for overall benefit to the semiconductor manufacturing ecosystem?

4. How do we expect supply chain evolution to progress in the face of the macroeconomic and geopolitical trends we are witnessing?

See Beyond the Horizon

www.linx-consulting.com

Summary

- Demand is strong
- Supply is tight
  - What could possibly go wrong?

- Global macroeconomic and geopolitical environments present challenges and opportunities
- The weather forecast is becoming more and more important
- Technology landscape is full of innovations and challenges
  - Quality (purity viz. particles & other FM)
  - Maintaining quality across complicated and long supply chains
  - Technology innovation
  - Process Changes & Change Control

See Beyond the Horizon

www.linx-consulting.com

Future of Materials: A Holistic Perspective

James A. O'Neill, Ph.D.
Senior Vice President and
Chief Technology Officer

Entegris

1

Long-term Industry Growth

Data Economy | IoT and Smart Devices | EV and New Energy | AI | Industrial Automation | Automotive and Autonomous Drive | VR and AR | Blockchain | 5G Network

By 2025 there will be over 75 billion installed IoT devices worldwide[1]
Volume of global data to increase 10× to 175 zettabytes from 2016 to 2025[2]

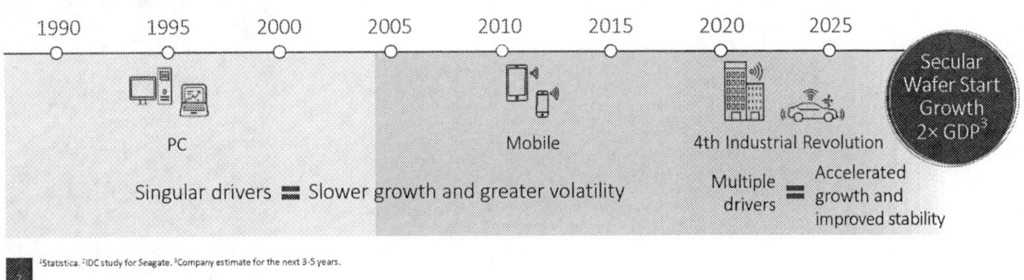

[1]Statistica. [2]IDC study for Seagate. [3]Company estimate for the next 3-5 years.

2

Entegris

Delivering Performance: Era of Increasing Materials Intensity

Entegris

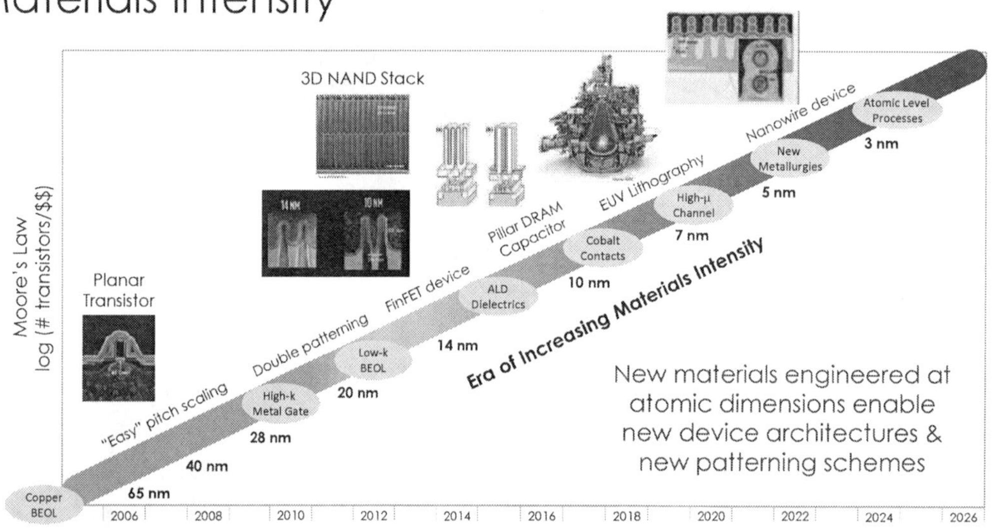

Higher Performance and More Complex Chips

Advanced nodes use more materials and present greater yield challenges

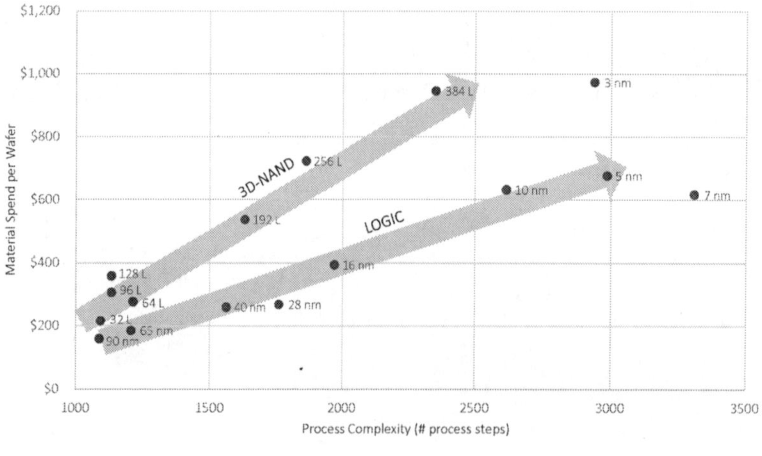

Source: IC Knowledge, Entegris.

3D-NAND
From 64 to 128 layers

CVD STEPS +48%

WET STEPS +36%

Logic
From 10 nm to 5 nm

ALD STEPS +33%

WET STEPS +6%

Entegris

The challenge our customers are facing

Entegris

More materials

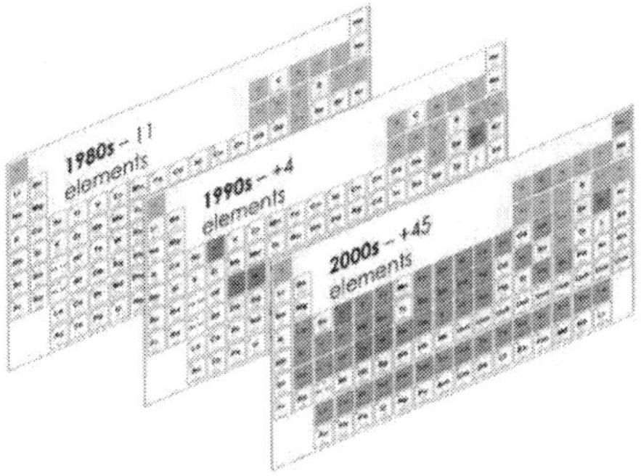

On more complicated geometries

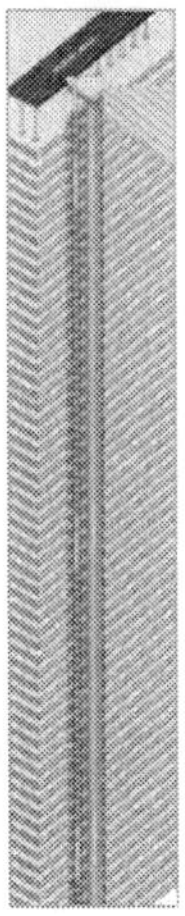

3D NAND Device

At smaller dimensions

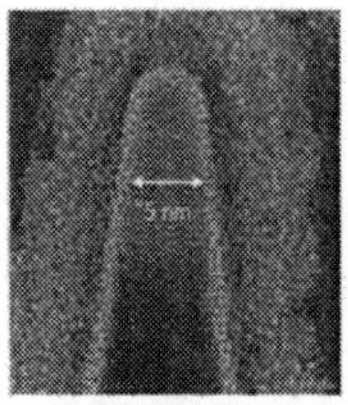

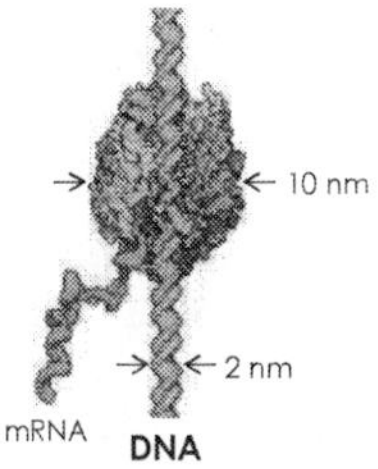

Interplay of materials on complex geometries at atomic scale

- Unique materials performance requirements
- Rigorous defect control (PPQ impurities, zero defects)

5

New device architectures and new materials introduce new and complex challenges.

Entegris

Increased Process Complexity

Change from 20 nm to 7 nm logic technology results in a **2x increase** in processing steps.

Faster Cycle Time

Production ramp for 10 nm foundry wafers is **6 months faster** than 16 nm wafers.

Advanced chips require **more materials** which means more process complexity and more challenging yield ramp.

Still, "speed to yield" remains critical for timely technology introduction.

Contamination control remains the largest challenge as wafer technology advances.

From 28 to 7 nm logic:

- Metal impurity concentration limit is reduced **1000x**
- Wafer killer particle size shrinks nearly **4x**

A **1% yield improvement** can mean:

$150M/year net profit in a leading-edge logic fab[1]

$110M/year net profit for 3D NAND fabs[2]

[1]Assumes 10 nm logic fab, 60K wafer starts per month
[2]Assumes 64 layer 3D NAND fab, 60K wafer starts per month

Entegris

7

How do we achieve ever more stringent purity requirements?

Entegris

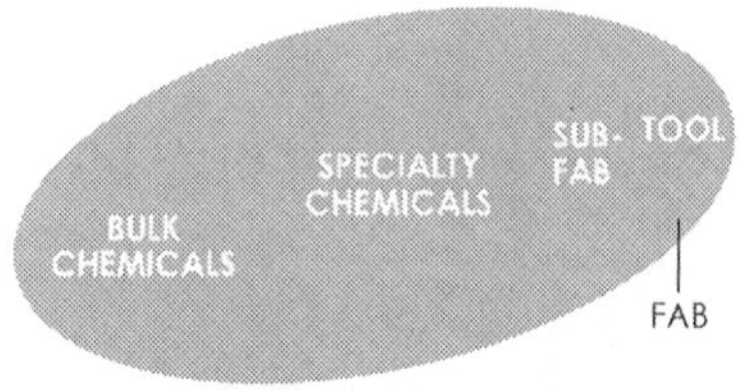

	1990-2000	2000-2010	2010-2020	2020-2030
Process node	500 nm	90 nm	10 nm	3 nm
Critical particle size	250 nm	45 nm	5 nm	1 nm
Metal purity	Not required	ppm	ppt	ppq
Filter retention size	200 nm	20 nm	3 nm	< 1 nm
Key product platform	Fluorogard®	Quick Change®	Torrento®	Future products
Filtration points	Tool	Tool, sub-fab	Tool, sub-fab, specialty chemicals	Whole ecosystem
Technology	Sieving	Sieving	Sieving, non-sieving	Sieving, non-sieving, molecular filtration

Over time contamination control requirements have expanded from the fab to the entire ecosystem → Need greater coordination, collaboration and drive

8

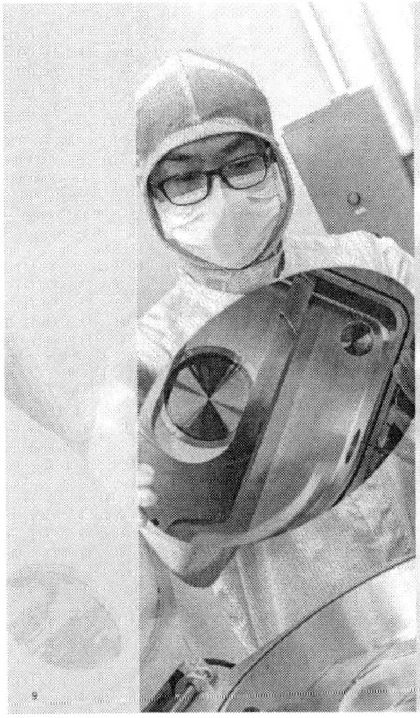

Filtration and Purification

- Process contamination is comprised of not only particles but also organics, metals, anions, moisture, and non-volatile residues
- **Contamination** can occur **at any point in the liquid, gas and air delivery systems** and at any point in the supply chain
- Filtration/Purification **solutions** are **not commodities** anymore – They are true **enablers** of the SEMI manufacturing process
- **Selection** of the **right solution** is complex and often needs to be **tailored** to the chemistry used in the application

Entegris

9

Common Challenges in Wet Processing As Scaling Continues

Entegris

Wetting and kinetics for nanostructured surfaces

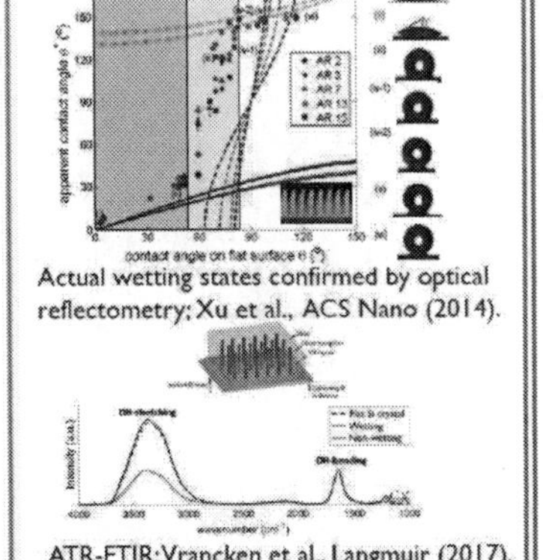

Actual wetting states confirmed by optical reflectometry; Xu et al., ACS Nano (2014).

ATR-FTIR; Vrancken et al., Langmuir (2017).

3D structures with more surface area → interfacial phenomena

Capillary force induced pattern collapse

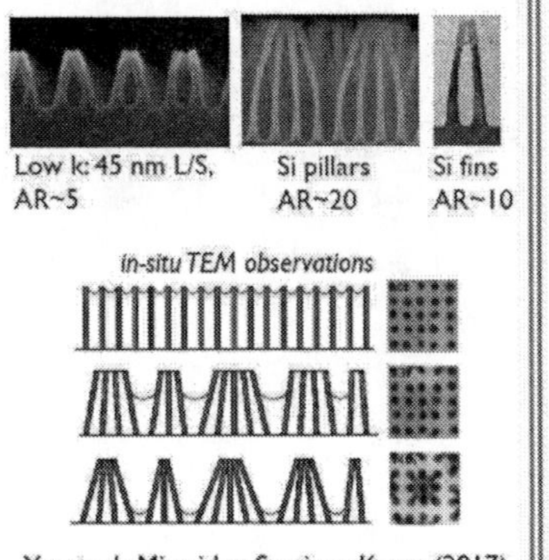

Xu et al., Mirsaidov, Semicon Korea (2017).

High surface to volume ratio → mechanical stability and structural integrity

Selective etch requirements

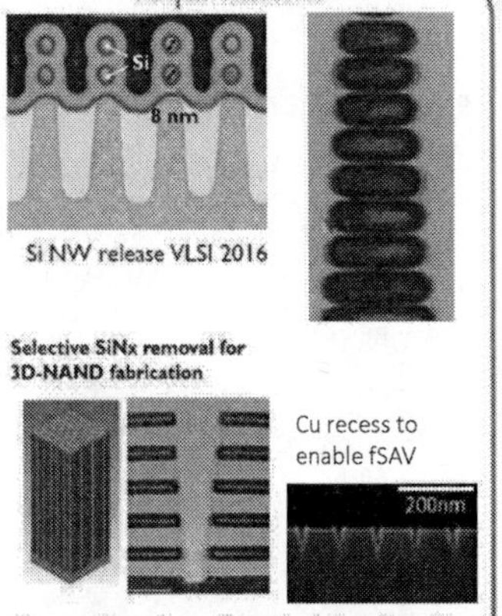

dimensional scaling → introduction of new device architectures and new selective etch requirements

imec

8

Atomic Scale Implications

Modeling Nano-scale Effects in Chemical Formulations

Entegris

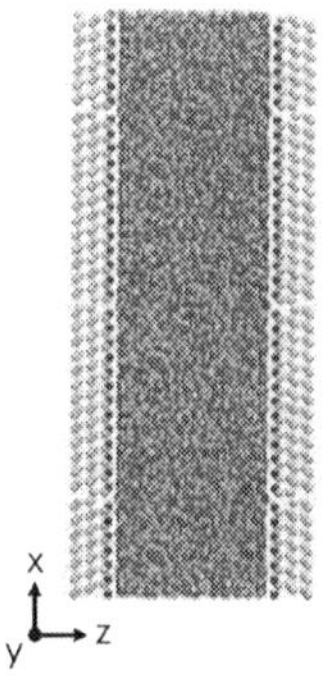

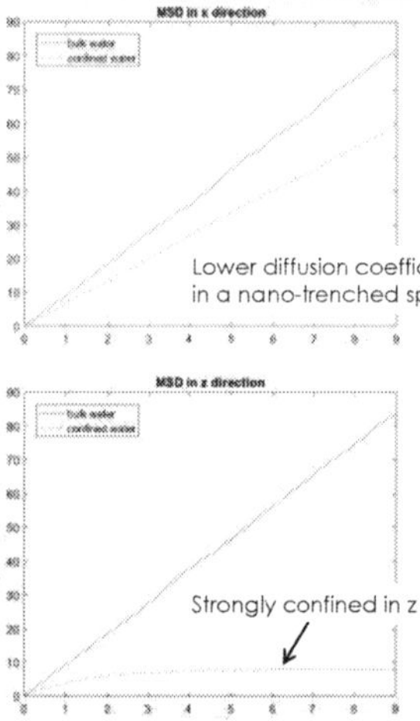

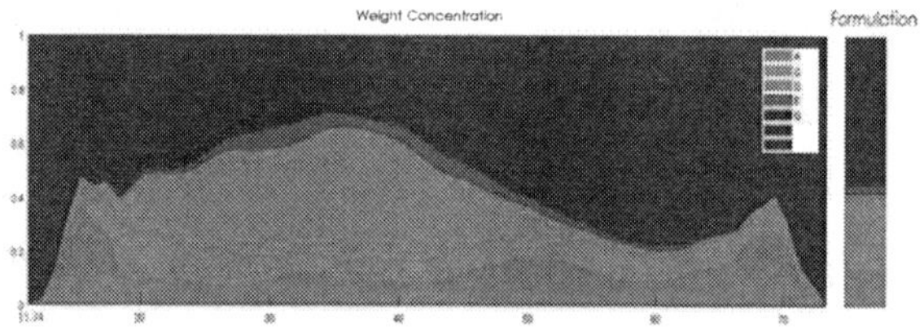

Component concentrations in nano-confined spaces different than in bulk formulation

11

Vertical Integration

Key challenges

- Increasingly narrow structures (96 L ⟶ >200 L)
- Very small features (atomic-scale processing)
- Complex material stack (high etch selectivity)
- Improved resistivity (new metallurgies)
- Improved defect control (extreme conformality)

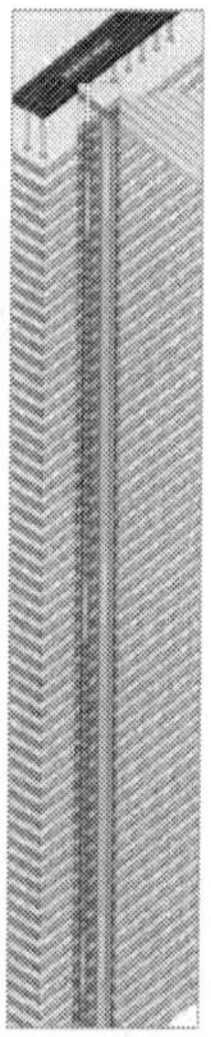

3D-NAND structure

High Aspect Ratio Ecosystem

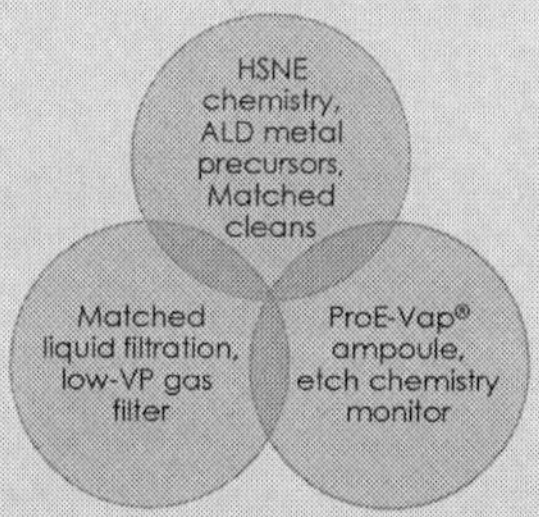

Unseen implications of engineering materials in tight spaces

Entegris

12

Post Cu/W Interconnects

Key challenges:

- Scalable interconnects (finer patterns)
- Improved resistivity and reliability (new metals)
- Downstream integration (CMP and wet processing)
- Improved defectivity control

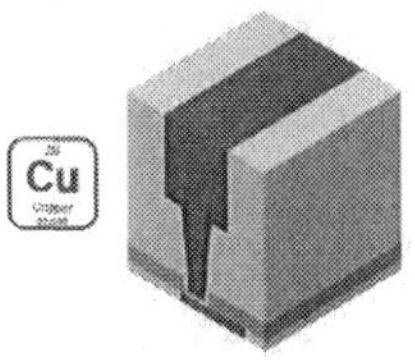

Copper interconnects require barrier layer that limits scaling

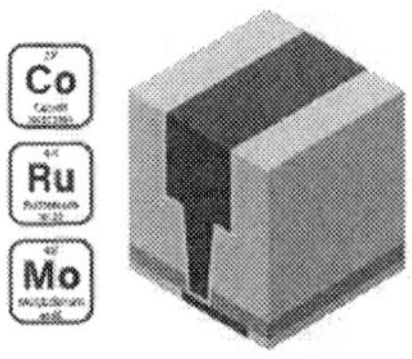

New materials required to scale advanced interconnect

New Metallurgy Ecosystem

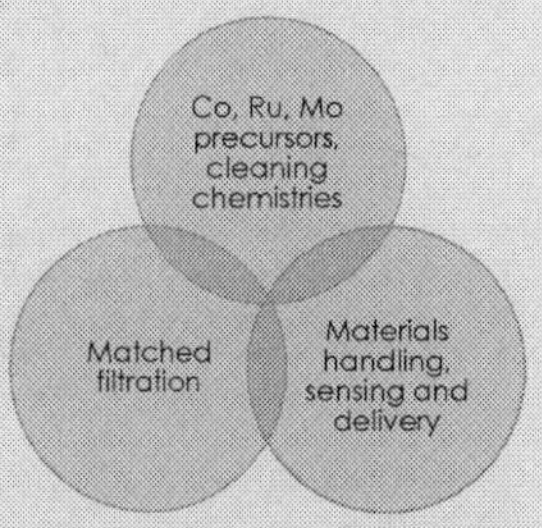

Cascade of implications resulting from simple change of materials

Entegris

13

Drive Toward Zero Defects

Key challenges:

- Increasing reliability requirements (automotive)
- Increasing process complexity (yield ramps)
- Undetectable defects (latent fails)
- Time-to-yield

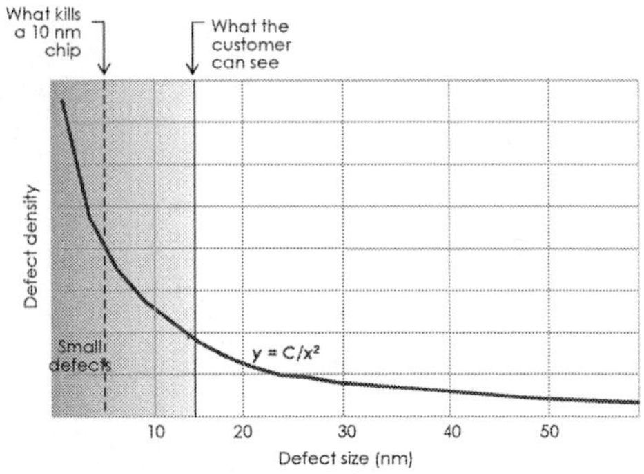

Customers increasingly requesting aggressive filtration as insurance against defects they cannot see

Zero Defect Ecosystem

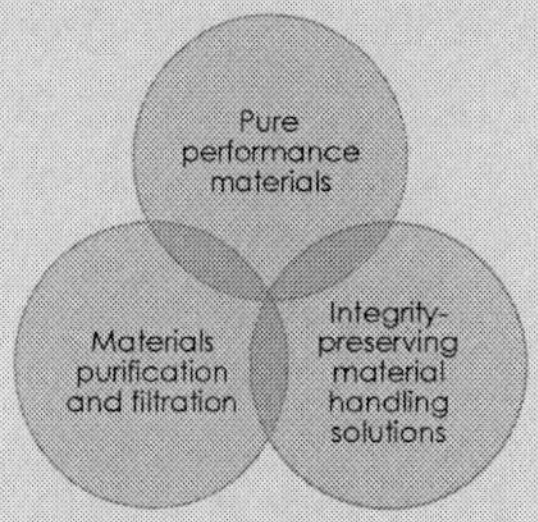

Material integrity through the Supply Chain

Entegris

14

Enabling Better Performance and Better Yield The Holistic Approach

Entegris

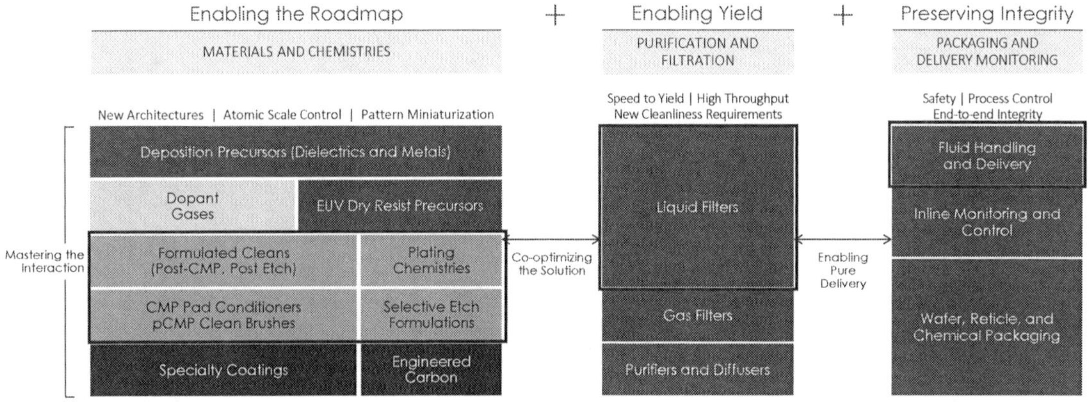

15

Leveraging the Whole Portfolio: New Metallurgies for 3D-NAND

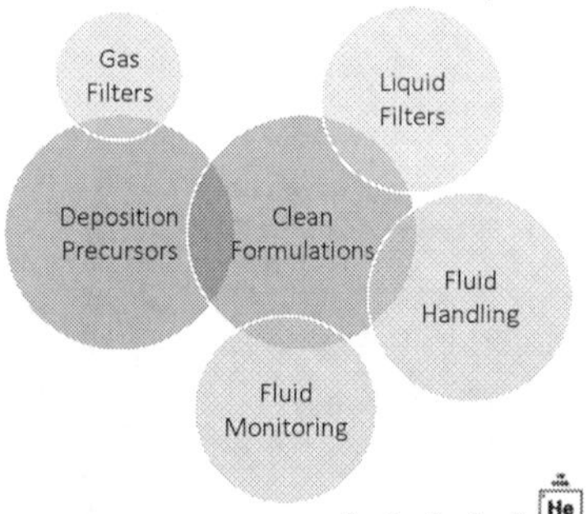

Cascade of value

Taller 3D-NAND requires more conductive metal (Mo)

New clean formulations required to match metallurgy

New filters required to match clean formulation

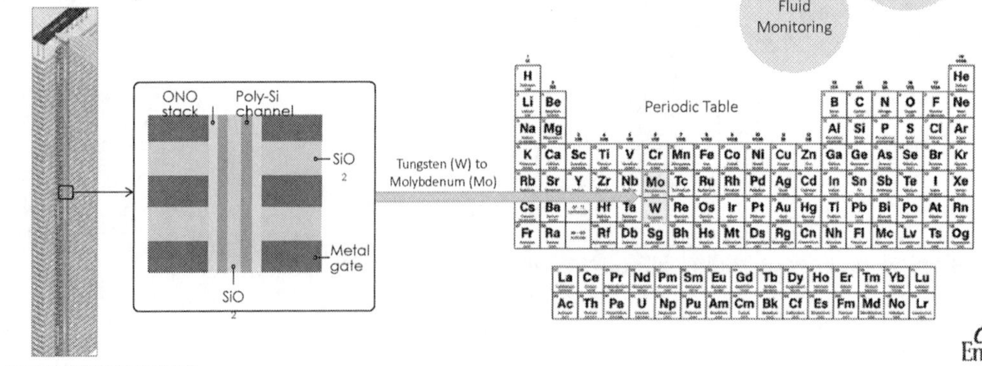

ONO – oxide-nitride-oxide dielectric stack.

Entegris

Engineering Selective Etch Solutions to Enable New Materials

Goal

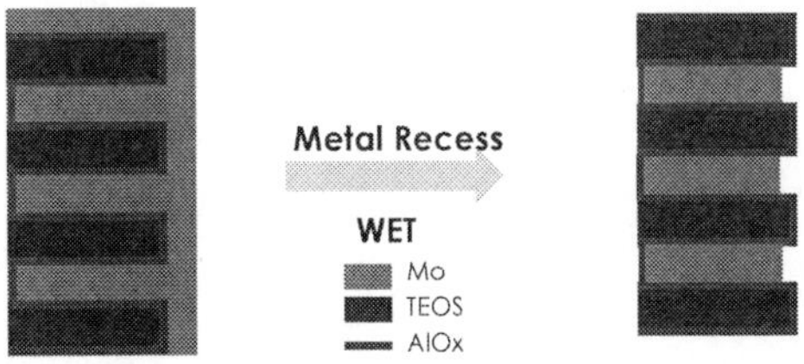

Effect of unmatched filters on etch formulation

Non-optimized filter can remove critical components of formulation and drastically affect etch performance

Challenges

- Additives tuned to achieve performance
- Filter tuned to match chemistry
- Monitor ensures process consistency

Entegris

Advanced Materials Learning

Advanced Materials learning is not just focused on the formulation

- Characterization of packaging interactions
- Consistent approach to material compatibility studies reduces end-product variation (material, packaging, and process equipment)

Material Learning / Process Optimization Alignment (dynamic relationship)

- Trade-offs between reducing material variation and understanding what leads to the most cost-effective processes for (addressing what matters)
- Early understanding of process windows and relationship to process parameter variation is key
  - Cost, capacity, and vertical supply integration

Entegris

Entegris
No matter what challenge comes next...
Entegris is committed to adaptive innovation.
Materials Innovation | Holistic Solutions | Strategic Partnerships
"The best way to predict the future is to invent it." – Alan Kay, 1940

Enabling Sustainable Semiconductor Manufacturing Applying First Principles to Contamination Control and Site Water Management

Slava Libman, Ph.D., PE

6/7/2021

FTD solutions

Semiconductor Industry Needs in UPW and Water Management during Turbulent Times

- Yield Enhancement via Technology Risk Management
- Water Management to Support Resilient Infrastructure
- Reduction of the Environmental footprint to support business sustainability
- Talent development to support business sustainability

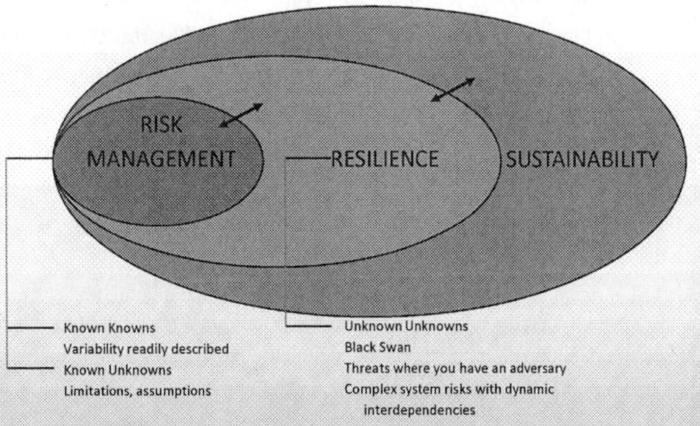

Source: UPM 2020 Panel Discussion

2

Technology Drivers, Needs, Challenges, Potential Solutions

Reliability in focus...

3

Cloud and edge computing drive More Moore

- Device-interconnect-memory technologies for mobile and cloud+HPC computing
- Mobile computing - additional functionality, biometrics, and display/camera/sensing for increased consumer value
- Cloud+HPC computing - 2.5D/3D integration to scale memory bandwidth / power and latency

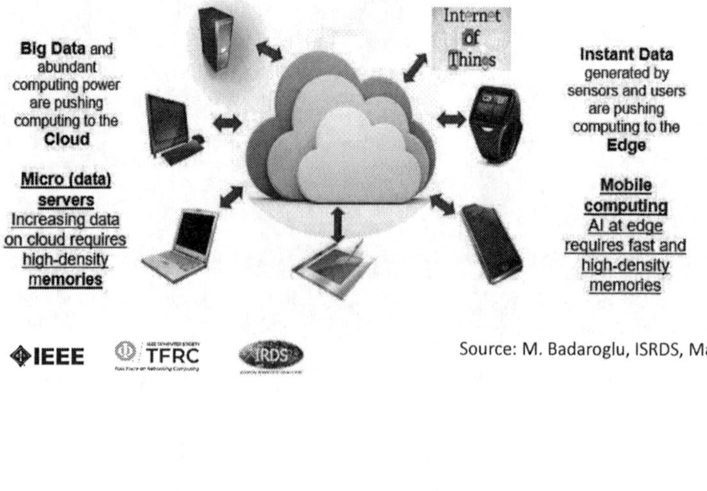

Source: M. Badaroglu, ISRDS, May, 2021

Dimensional Scaling Expected to Stall >2028

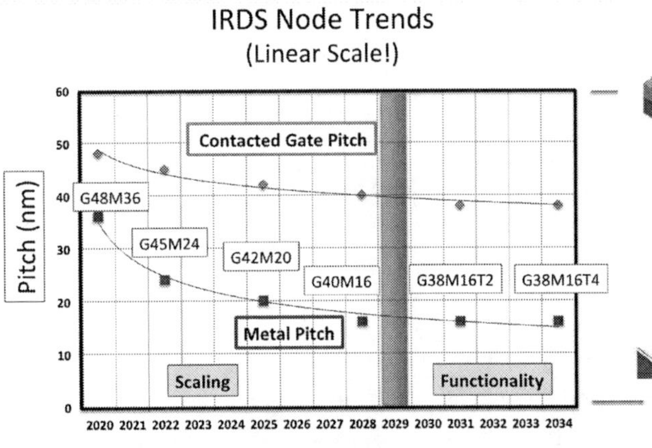

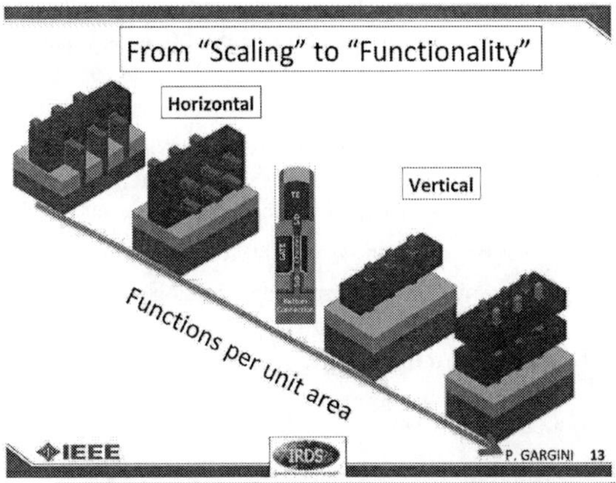

5

Defect control – driven by unit processes

- Deposition and etchant materials less likely to change but
  - Non-Cu metallization – Co, Ru
  - New di-electric systems for low-k spacer and self-deposition
  - Si-SiGe heterostructures for nanosheet formation
  - Zero-tolerance to defects in hybrid bonding
- Dimensions getting smaller and higher aspect ratios
  - Taller fins and gates
  - Vertical device
  - More need for conformal deposition and sidewall/bottom cleans
- More material systems and complex structures during deposition & etch
  - Complexity of local interconnect
  - Complex etch selectivity matrix
  - More surface interactions
  - Sidewall damage control
  - More critical loop processes as all cannot be done in-situ
- More reliance on self-alignment and self-deposition
  - Multi-material systems
  - Electrical isolation risks

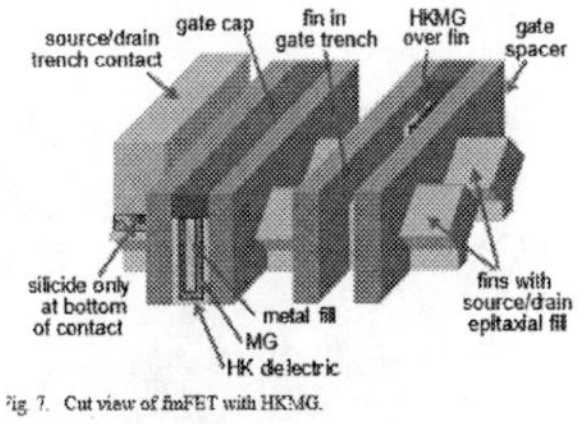

Fig 7. Cut view of finFET with HKMG.

Source: Qualcomm@CICC 2018

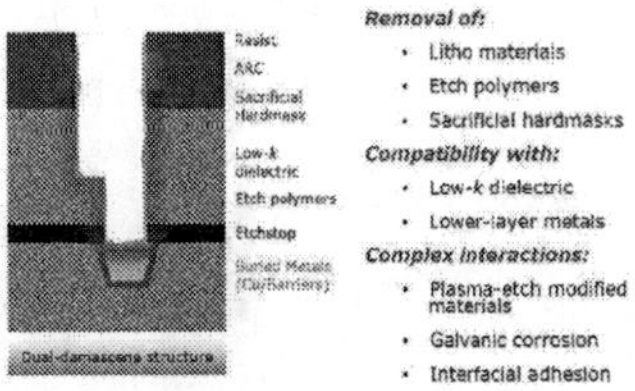

Source: Intel@ECS 2011

IEEE

Reliability and yield – product lifetime implications

- Complex chips tightening the D0 requirements
  - More process steps
  - Larger Si chips (50-150mm2 for high-end AI inference, >500mm2 for AI training) versus 70-80mm2 earlier mobile chips
- Stringent mission profiles and operating range
  - >5 years for server chips from >115C degrees
  - >10 years for automotive chips from -40C to 125C
- BTI, HCI, and self-heating getting important
- Critical defect sizes are at the single-nm scale
- Correlation between yield and lifetime
  - Systematic defects
  - Random defects

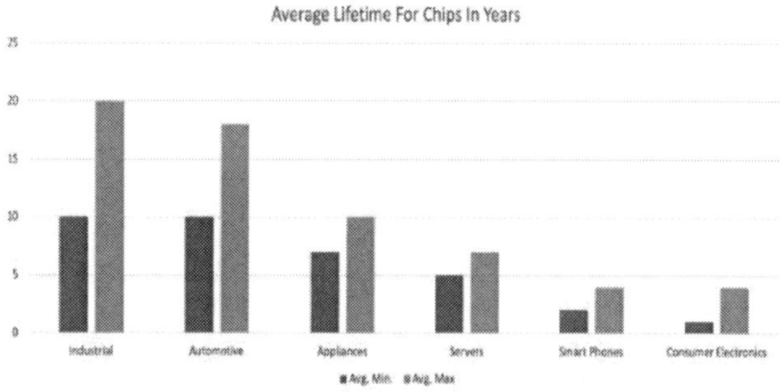

Source: M. Badaroglu, ISRDS, May, 2021

Yield Enhancement – Yearly Process Cycle

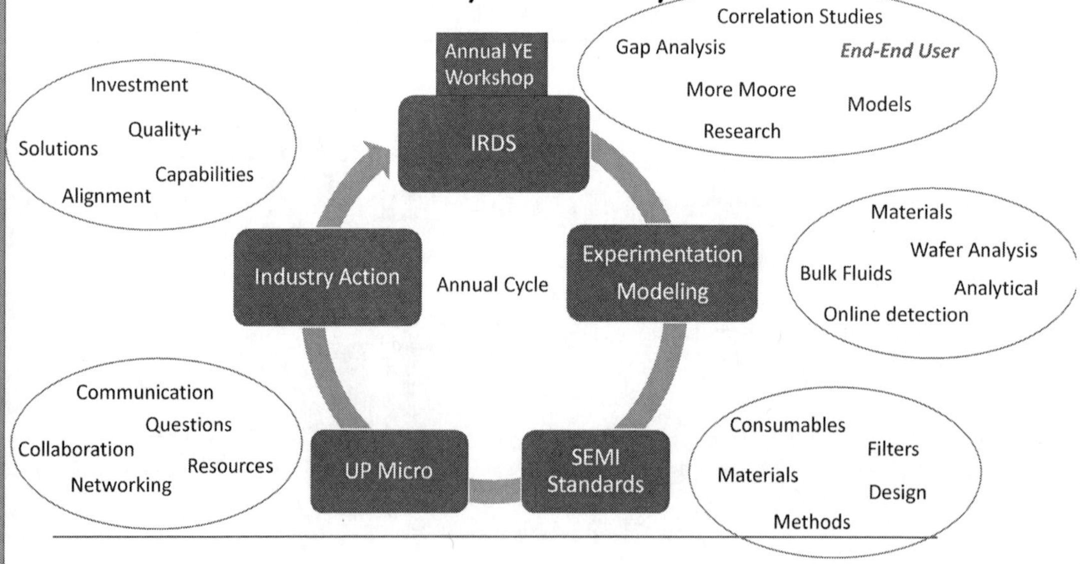

8

Examples of IRDS Development: Particles and Particle Precursors.....

9

UPW Particles – High Risk that Needs Action

- No effective metrology exists to accurately monitor sub-10nm particles
- Filtration capability at the killer particles size is marginal
- High Molecular Polymers may become killer particles
- Colloidal Silica Needs to be effectively controlled
  - Most difficult to retain
  - Typically occurring in UPW

Defect Density

$$Y_r = \frac{1}{(1 + AD_0)^n}$$

2020	CRITICAL AREA						DEFECT DENSITY	
	Overall Complexity	Width (nm)	Defect Size (nm)	Pitch (nm)	Density@Chip	Critical Area (cm2) in 80mm2	Dx	Dx/wafer
Gate	0.74	20	10.0	48	33%	0.267	0.0081	5.7
Fin	0.74	7	3.5	28	10%	0.080	0.1889	133.5
VC	1.48	16	8.0	48	2%	0.017	0.0158	11.2
MetalC	2.23	16	8.0	48	27%	0.213	0.0158	11.2
Via0	0.74	15	7.5	42	2%	0.017	0.0192	13.6
Metal0	3.71	15	7.5	30	40%	0.320	0.0192	13.6
Viax	1.48	18	9.0	51	1%	0.010	0.0111	7.9
Metalx	4.45	18	9.0	36	40%	0.320	0.0111	7.9
Viay	2.23	40	20.0	113	1%	0.005	0.0010	0.7
Metaly	2.23	40	20.0	80	40%	0.320	0.0010	0.7

Particles Targets

11

Concern of the particle size dependency

Particle; SiO2 particles
Particle size; 30~500 nm
Concentration; 10, 100 #/mL
Pretreatment; O3 rinse
Procedure; Rinse w/particle 60-1020 sec. → Dry 30sec.
Wet Tool: SU3200

Previous approach assumed no particles dependency.... Is this the case?

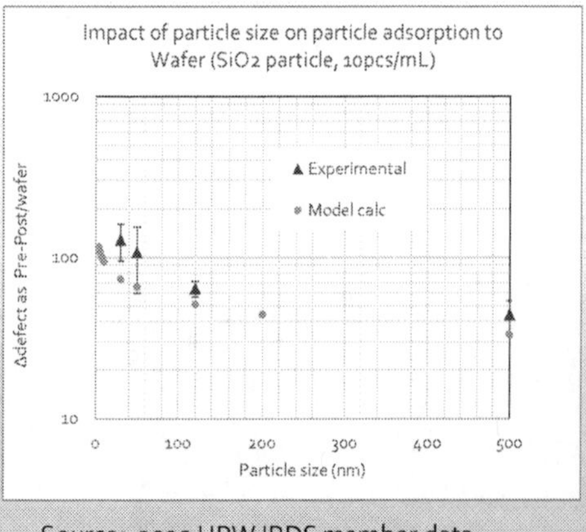

Source: 2020 UPW IRDS member data

Quantification of liquid particle concentration

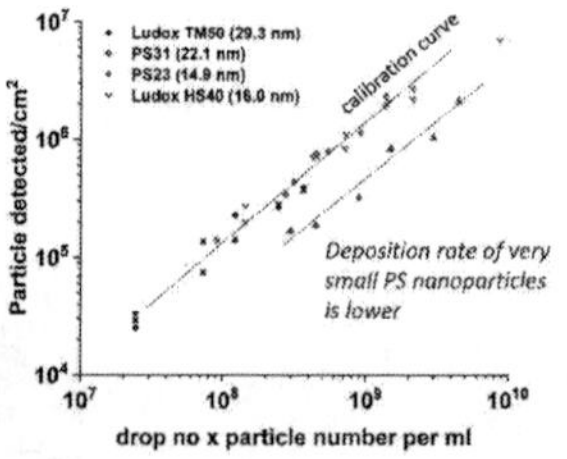

Drop number	Time required for successive spin-drying	Particle concentration in UPW [particles/ml]	Detected particles [per cm2]
1	1 min	1'000	1
10	2 min	100	1
100	8 min	10	1
1'000*	70 min	1	1

*Extrapolated data

Source: SPCC 2020

12

UPW IRDS – Particle Deposition Test

- Spin Coater was provided to CTA by Entegris
- Intel and Pall covered some of the cost

- Particle size from 5nm to 70nm
- Liquid concentrations for spin-drying:
  - Wafer-1: 1E9/ml
  - Wafer-2: 1E7/ml
  - Wafer-3: 2E6/ml

Sample preparation @ CT Associates

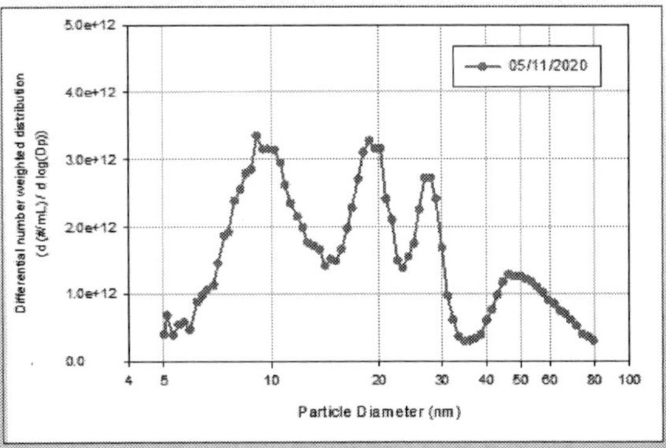

Wafer-scanning @ UNISERS

Shipped to Zurich

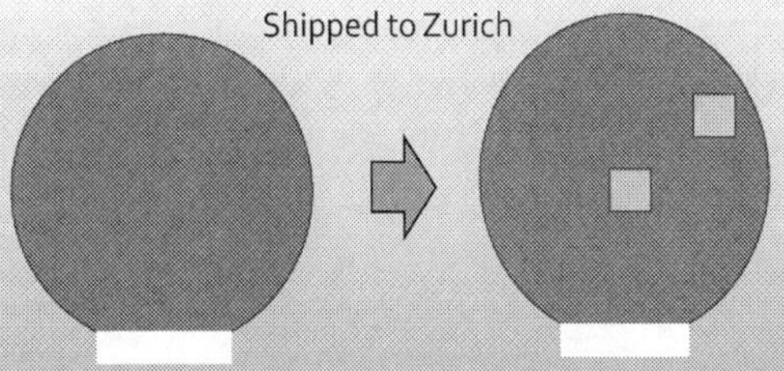

1cmx1cm scan @center & edge of the wafers

UNISERS: WAFER PARTICLE ANALYSIS

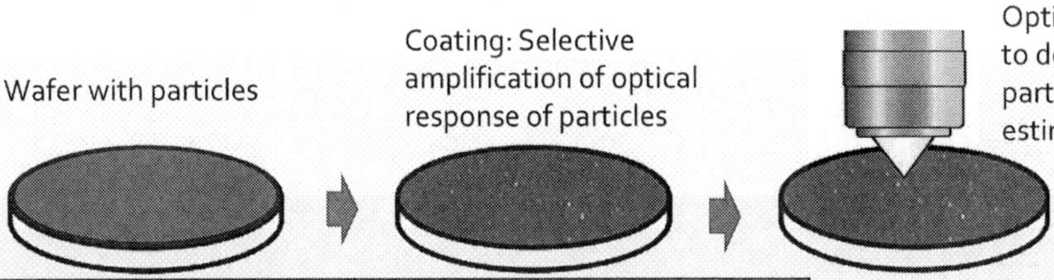

Defect detection limit on polished silicon	10nm
Defect detection limit on metallic films	10nm
Defect speciation* limit	(ongoing research) ~50nm
Defect size estimation	True size estimation (not PS equivalent)
Wafer size / type	100mm / Si, SiO2, Metallic, etc.
Automation level	Manual wafer handling
Disadvantage	Initial coating for signal enhancement required.

*Speciation is based on enhanced Raman (molecular spectroscopy). Strong technique for organics speciation, applicable for dielectrics (or metal oxide) speciation. Can detect a defect as "metallic cluster", but can not make metallic speciation.

RESULTS

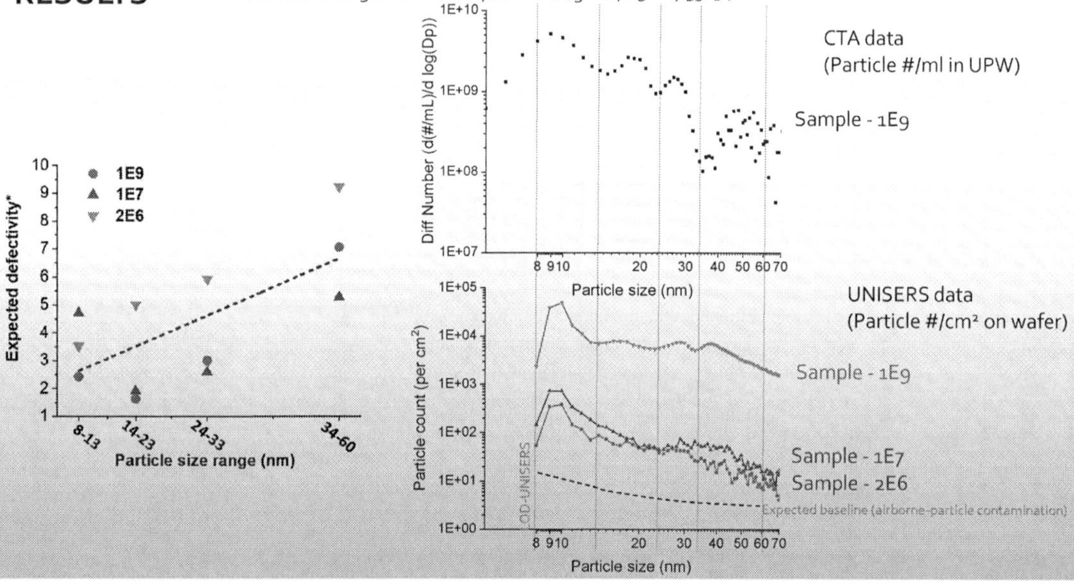

Particle Deposition Rate

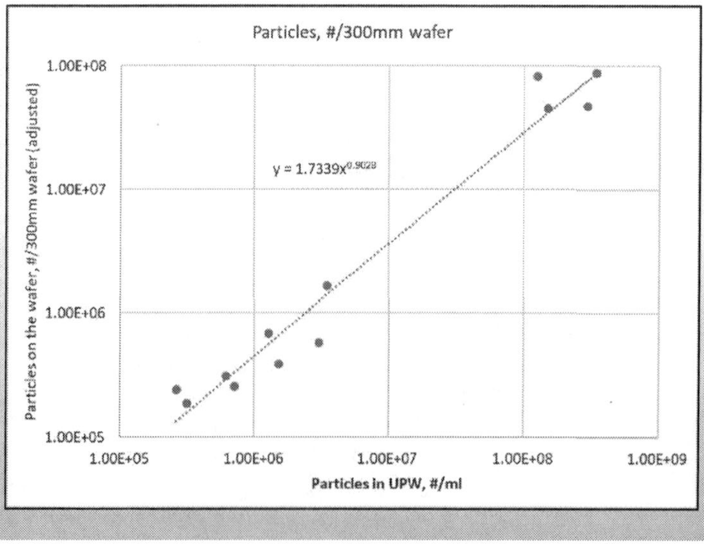

16

2021 YE Roadmap Updates – Particle Focus

Table YE3 Technology Requirements for Wafer Environmental Contamination Control

Year of Production	2021	2022	2023	2024	2025	2026	2027
Add Pitch designation							
Add Pitch designation							
Logic industry "Node Range" Labeling (nm)	"5"	"3"	"3"	"3"	"2.1"	"2.1"	"2.1"
IDM-Foundry node labeling	i7-f5	i3-f2.1	i3-f2.1	i3-f2.1	i2.1-f1.5	i2.1-f1.5	i2.1-f1.5
Logic device structure options	FinFet	finFET LGAA	finFET LGAA	finFET LGAA	LGAA	LGAA	LGAA
Logic device mainstream device	FinFet	finFET	finFET	finFET	LGAA	LGAA	LGAA
MPU/SoC Metalx ½ Pitch (nm)[1, 2]	12.0	12.0	12.0	12.0	12.0	12.0	12.0
Critical particle size non-electrically active (non-EAP) (nm) based on 50% of Logic 1/2 Pitch (nm (contacted) [1]	10	9	9	9	7	7	7
Critical particle size (nm) of Electrically Active particles based on 50% width of fin Logic SiGe Front End or other device critical dimensions for LGAA (>2 monolayers)	3.5	3	3	3	3.5	3.5	3.5
Critical size (EUV mask), nm	12	12	12	12	12	12	12
Critical particle size non-electrically active (non-EAP) (nm) based on 50% of Logic 1/2 Pitch (nm (contacted) [1]	10	9	9	9	7	7	7
Number of non-EAP at critical size (#/ml) [26], POE	1	1	1	1	1	1	1
Critical particle size (nm) of Electrically Active particles based on 50% width of fin Logic SiGe Front End or other device critical dimensions for LGAA (>2 monolayers)	3.5	3	3	3	3.5	3.5	3.5
Number of EAP at critical particle size (#/ml) [26], POP	100	100	100	100	10	10	10
Critical size (EUV mask substrate), nm	12	12	12	12	12	12	12
Number of particles for EUV mask production (#/ml) [26.3], POE	1	1	1	1	1	1	1
PROACTIVE: ~~EAP~~ Particle Control: 50nm (#/L) [26], Feed to the Final Filter. Values with (*) require particle control at 30nm size and larger. [refer to the Particle White Paper]	<50	<50	<50	<50	<50*	<50*	<50*
Particle Precursors, #/ml [56]	1 (see note)	1 (see note)	1 (see note)	1 (see note)	1 (see note)	1 (see note)	1 (see note)

POTENTIAL SOLUTIONS

Proactive Technology Management

Reactive

Measure
- Wafer Defect
- Final Quality Spec Parameter

Control
- Find Cause and Eliminate
- New Solution under pressure

Roadmap
- Define Spec Parameter
- Do nothing when cannot be reached

Proactive

Measure
- Everywhere feasible and critical
- New Spec Parameter (upstream, close to the contamination source)

Control
- Proactively Manage Causes
- Continuous Improvement

Roadmap
- Define New Spec Parameter
- Tight QA/QC throughout Supply Chain

Proactive

Proactive Technology Management

- Risk analysis
- Experimental validation of the risks
- Benchmarking and Roadmap Studies to help to commercialize solutions

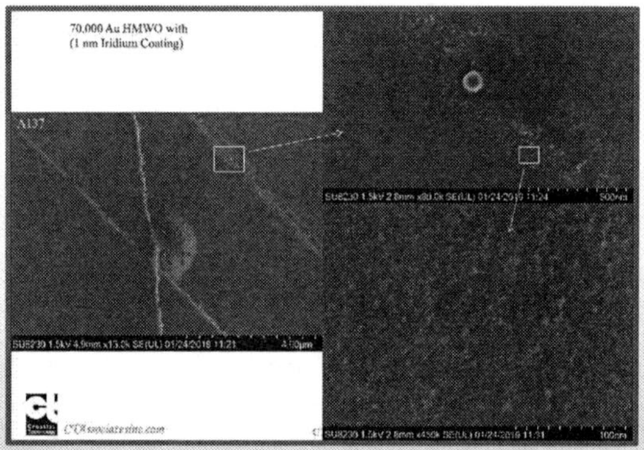

Example:
High molecular weight organics:
– Poly(sodium 4-styrenesulfonate) – Ave Mw ~ 70,000 da + 100 nm silica tracer
Media – epi-ready 1" silicon wafer

Defect/particle size distribution (Resin Extract)

8nm-19nm particle count (/ml): 7.55 E5
8nm-19nm defect count (/cm2): 1.58 E4

Slot ID-22 Resin-extract 1:600

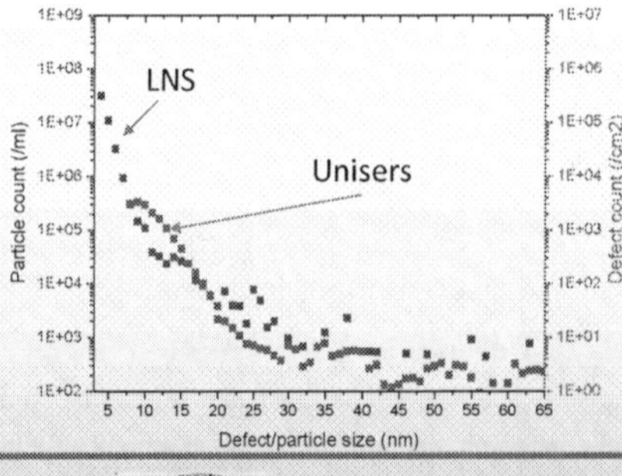

8nm-19nm particle count (/ml): 1.51 E5
8nm-19nm defect count (/cm2): 3.49 E3

Slot ID-17 Resin-extract 1:3000

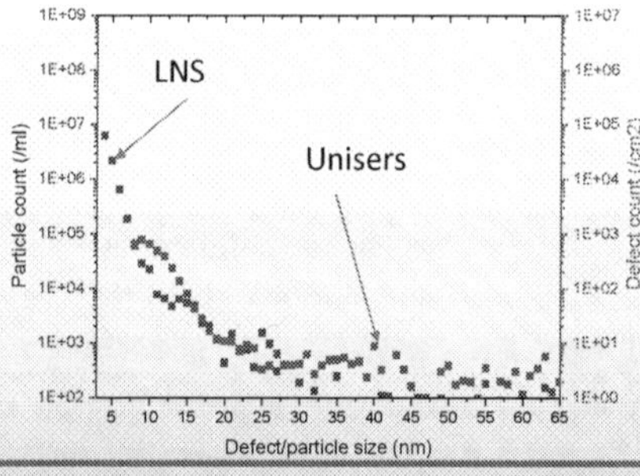

SP5 scan

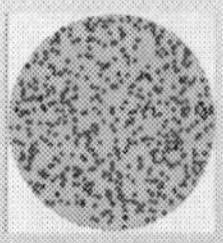

873 added (>19nm)

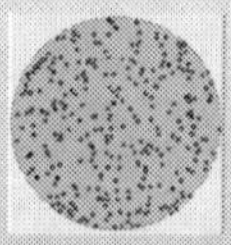

480 added (>19nm)

Source: UPW IRDS Study 2020

SEMI STANDARDS - Industry Support

- *Do we need more?*

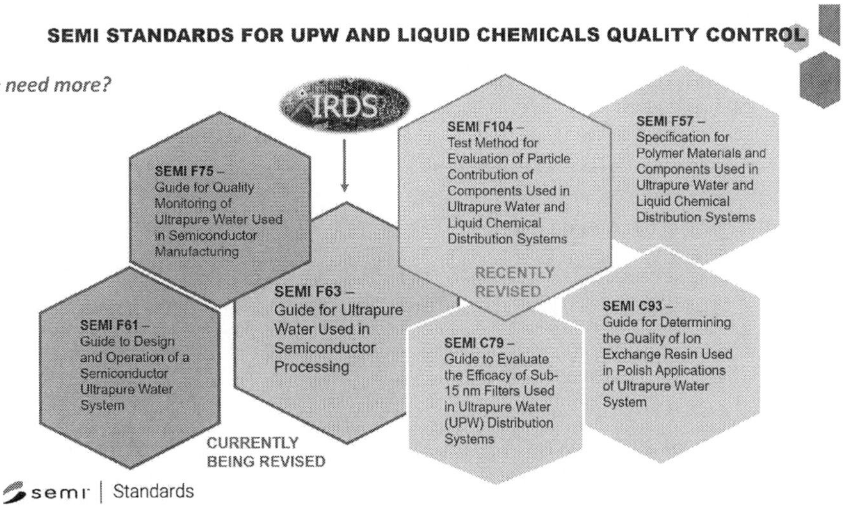

Site Water Management

22

Relationships Between Purity and Infrastructure Needs

- High device complexity and HOC reliability requirements drive higher purity requirements
- Increasing environmental challenges drive comprehensive water management solutions pushing towards UPW recycling (reclaiming wastewater for UPW make-up)
- As the result Wastewater Reclamation solutions have to be reliable and highly robust, adding even higher complexity to the Site Water Management
  - Mega-site complex infrastructure
  - Chemically intensive manufacturing
  - Sensitivity to energy consumption
  - Growing environmental requirements

Needs for wholistic approach

23

Water Management Challenges

Growing complexity due to:
- increased use of chemical, energy, and water
- facilities are growing putting pressure on the external infrastructure
- environmental compliance requirements are getting tighter
- sustainability goals have become an important additional driver

Why Now?

24

Leveraging SEMI Standards to Address Water Management Challenges

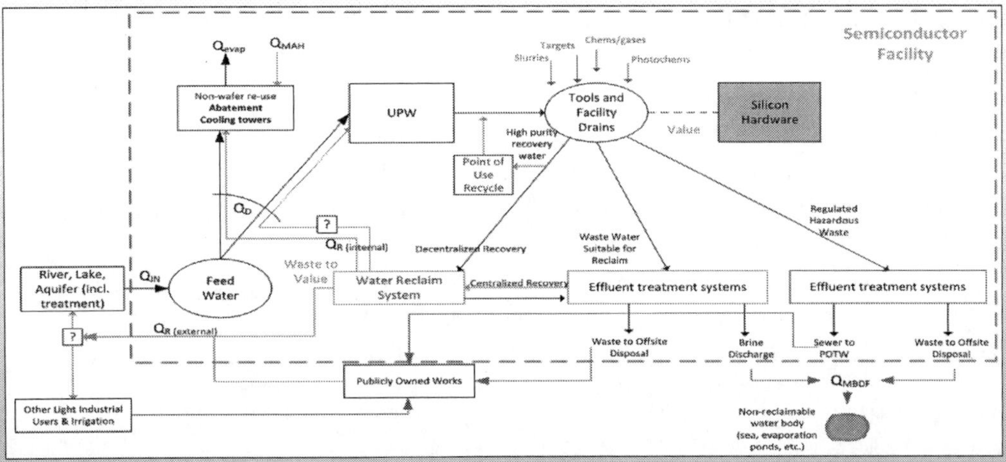

SEMI F98

25

Key for Water Management Sustainability

- Tight and Effective Data Management
  - Excessive amount of data generated, but not used
  - Quality of the data
  - Data integration
- Strategy - no one-fits-all
  - Site specific conditions
    - Climate
    - Regulations and infrastructure conditions and constraints
    - Time-to-market
    - Organization
- Water Management Roadmap
  - Planning
  - Streamlined decision

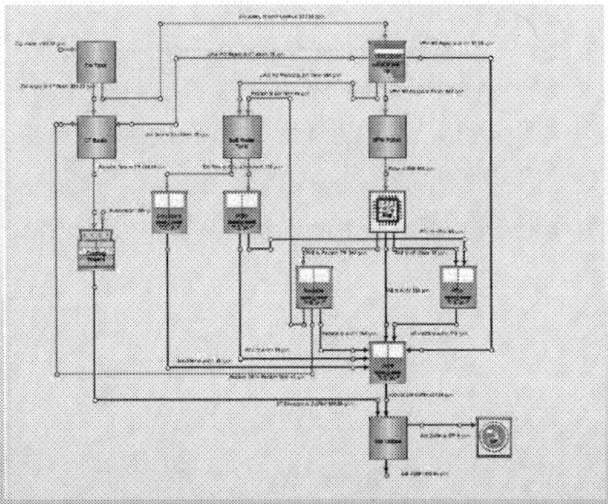

Summary

- Advanced existing and future semiconductor facilities must deal with complexity of enabling sustainability of both contamination control and infrastructure constraints.
  - Reliability on HPC is in focus of the technology roadmap
  - Particles and particle precursors are some of the critical issues
  - Recycling of UPW is the connection between contamination and infrastructure management issues
  - Integrated wholistic approach is needed to ensure long term sustainability
  - Proactive technology management throughout facility and its supply chain is need more than ever
    - Contamination control cannot rely on "reactive" approach due to lacking metrology
    - Site water management requires systematic assessment, creative strategy, and long range planning

27

Acknowledgement

- Excellent contribution of most YE IRDS and SEMI Liquid Chemical Committee
  - Micron, Intel, Samsung, GlobalFoundries, Kanomax-FMT, Screen, Pall, Evoqua, Suez, AMAT, Kurita, Purdue, +GF+, Entegris, Lanxess, Balazs, 3M, FTD Solutions
- Special thanks for Samsung, CTA, and Unisers who conducted the experimental work
- IEEE IRDS, Semi, and GWI who enable industry wide collaboration

Questions?

Slava Libman, PhD, PE

FTD solutions

slibman@ftdsolutions.net

ftdsolutions.net

29

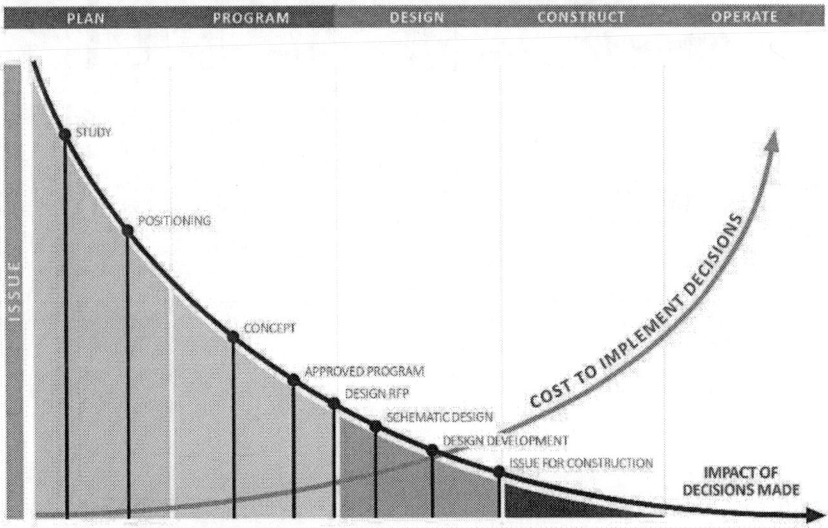

FTD Scope

30

Surfactant rinse solution for EUV photo-lithographic process

PN Liao, Mira Park, Johnny Tsai, Joyce Chin, Ping Hsu
Advanced Clean Technology, DuPont
Shared Lab, Interconnect Solutions, Dupont

1. Introduction

Photoresist rinse step is necessary after TMAH develop in photo-lithographic process to prevent photoresist pattern collapse and clean polymer residues. Unbalanced capillary force on resist side wall is the major cause of pattern collapse during spin dry process. Surfactant contained formulations are able to prevent resist collapse but the high affinity of surfactant molecules to photoresist is shown to result in CD-shift and even resist melting.

2. Fundamental

Unbalanced capillary force on resist side wall is the major cause of pattern collapse during spin dry process.

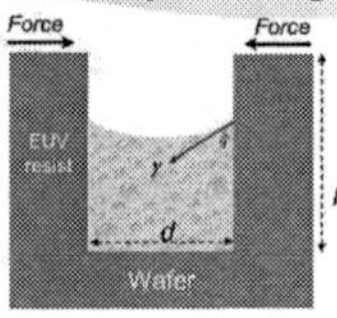

$$Capillary\ force = \frac{2\gamma LH\cos\theta}{d}$$

γ: surface tension
θ: contact angle
H: Height of PR
L: Length of PR
D: Width of PR

(i) Capillary force is determined by pattern dimension and solution physical properties. Contact angle on sidewall is one of the main factors of pattern collapse.

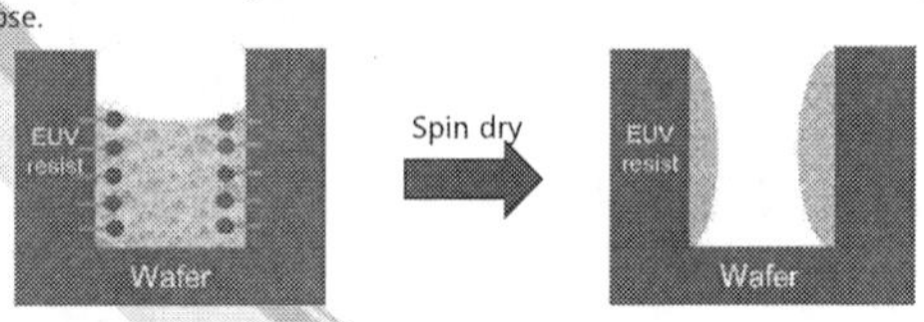

(ii) High affinity surfactant molecules penetrate into photoresist and cause PR swelling.

(iii) In order to study sidewall properties, a photoresist was partially light exposed. The sidewall is created after treated with TMAH solution to remove the top layer of deprotected photoresist.

3. Results and Discussion

➢ **Create a sidewall surface with a partial exposure photoresist**

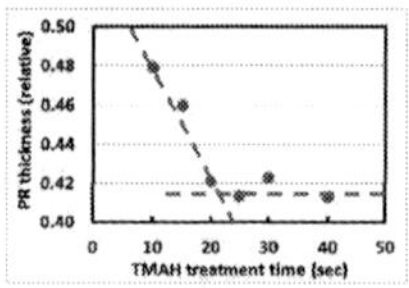

Fig. 1: Film thickness of partial exposure PR after dipping in TMAH 2.38% solution.

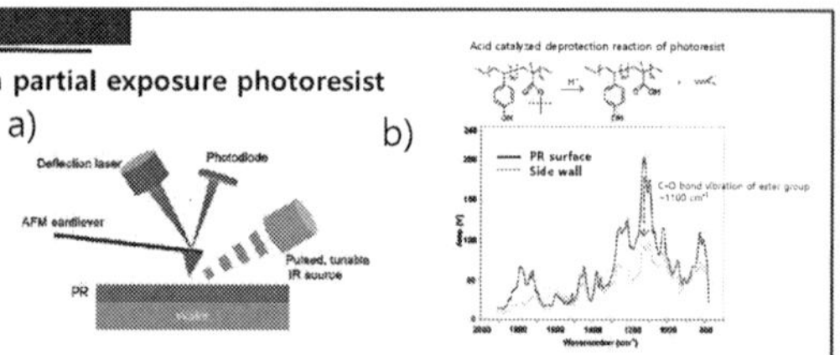

Fig. 2: a) Schematic diagram of AFM-IR. b) AFM-IR spectra of photoresist surface and side wall surface.

➢ **Determine contact angle of DIW and surfactant solution on sidewall and photoresist surface**

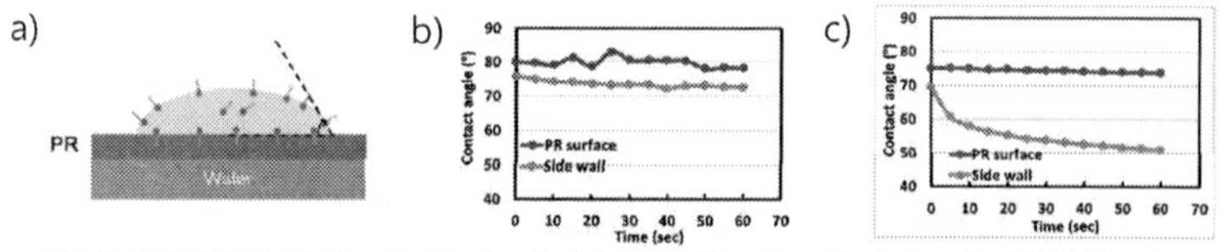

Fig. 3: a) Measurement of contact angle of solutions on PR surface. Contact angle of DIW (b) and a surfactant solution (c) on PR and sidewall surface.

➢ **Photoresist swelling in different surfactant solutions**

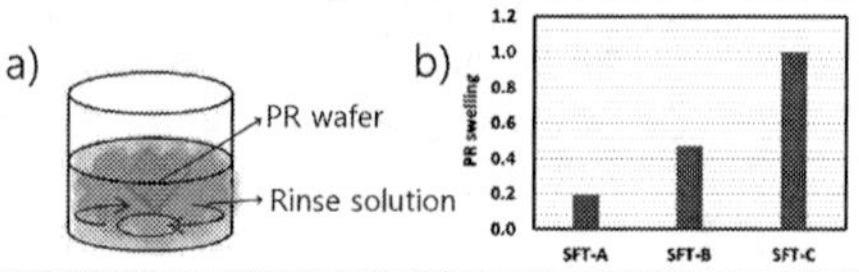

Fig 4: a) Partial exposure PR coupon dipping in surfactant solutions after TMAH pre-treatment to create sidewall surface. b) S urfactant solution A, B, and C are the same concentration but they cause different level of PR swelling.

4. Conclusions or Future Application

Photoresist rinse step is an important process to prevent pattern collapse and clean polymer residues. Contact angle on sidewall is one of the factors to prevent pattern collapse. Through our inhouse study, sidewall has different properties from photoresist surface. It is more hydrophilic and has lower contact angle with rinse solutions. High affinity surfactant molecules to photoresist resuls in PR swelling. 3 surfactants were used in this study and they cause different level of PR swelling.
With the understanding of interactions between surfactant and photoresist sidewall, we are able to create a rinse solution that has pattern collapse mitigation, good clean performance, and controlled CD shift.

ACT

pn.liao@dupont.com

Effect of power and surfactant on the behavior of bubbles in an ultrasonic field

Bichitra Nanda Sahoo[1], So Young Han[1], Nagendra Prasad Yerriboina[1], Keita Ando[2], Brown Kang[3], Andreas Klipp[4], and Jin-Goo Park[1*]
[1]Department of Materials Science and Chemical Engineering, Hanyang University, Ansan, 15588, Republic of Korea
[2]Department of Mechanical Engineering, Keio University, Yokohama 223-8522, Japan
[3]Electronic Material R&D Center Asia, BASF, Suwon, 16419, Republic of Korea
[4]BASF SE, Ludwigshafen, 67056, Germany
(*E-mail: jgpark@hanyang.ac.kr)

BASF
The Chemical Company

The surface preparation and cleaning conference (SPCC)-2021, June 8-10, 2021

Introduction

The main objective is to study the effect of ultrasonic power and concentration of surfactant (SDS) on the characteristics of artificially generated bubbles.

Experimental procedure

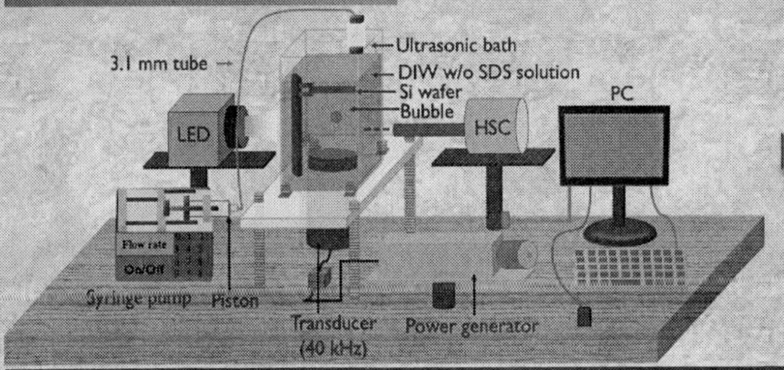

Results

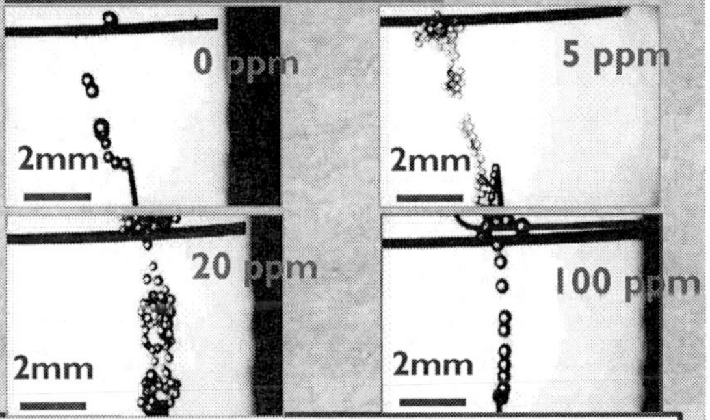

Cleaning solutions : DIW w/o SDS, **Flow rate of air** : 2ml/min, **Ultrasonic power** : 4 watt

Conclusion

- At low surfactant concentration, bubbles are vibrated due to high transient cavitation
- At 20 ppm of SDS surfactant, population density of artificial bubbles is maximum
- With increasing conc. of surfactant, transient cavitation reduced
- At 100 ppm concentration of SDS surfactant, no disturbance occurs with motion of bubbles.

Advanced Solutions for Improved Post-CMP Cleaning of Ceria-Based Residues

Katie Gramigna, Paul Bernatis, Wayne Huang, and Fengji Yeh
Advanced Cleans Technology, DuPont Electronics & Industrial; contact: katie.m.gramigna@dupont.com

Introduction

To achieve rapid removal rates during the chemical mechanical planarization (CMP) of silicon dioxide (TEOS), along with high selectivity for polishing TEOS over SiN_x, ceria-based slurries are employed in STI- and ILD-CMP, as well as in advanced memory applications. This polishing efficiency and selectivity is a result of two predominant properties: (1) the relatively strong bonding that forms between ceria and the TEOS surface, and (2) the strong electrostatic attraction between ceria particles (IEP ~6-8) and the TEOS surface (IEP ~2-3) under acidic conditions. However, these properties render byproducts from the CMP process particularly challenging to remove from the wafer surface, as the residual Si-O-Ce bonding must be chemically broken, and the electrostatic attraction between the ceria particle and the TEOS surface must be overcome.

Problem

Common post-CMP (PCMP) cleaning processes for the front-end-of-line (FEOL) include scrubbing the wafer with PVA roller brushes using commodity chemicals such as dilute HF, dilute NH_4OH, citric acid, NH_4OH/H_2O_2 mixture (SC-1), and TMAH. However, for cleaning difficult to remove particles, such as strongly adhered ceria-based residue, these cleans often provide insufficient ceria removal or lead to excessive material loss, so an H_2SO_4/H_2O_2 mixture (SPM) often supplements brush cleaning to enhance cerium ion removal. However, SPM is undesirable for industrial use owing to the associated safety and environmental concerns along with a high associated cost of ownership of the off-line process step. Therefore, the development of new strategies to improve cerium residue cleaning has been targeted by DuPont.

Overview of the performance of commodity chemical-based PCMP cleaners:

Commodity Cleaners	NH_4OH	TMAH	SC-1	SPM	HF
Dielectric compatibility	[illegible]	[illegible]	[illegible]	-	[illegible]
Cerium residue removal	X	X	X	O	X
Safety	[illegible]				
Direct clean	O	O	O	[illegible]	O

Materials and Background

Goal: To remove cerium residue remaining on wafer after CMP processing by dissolution

1) Develop cleaning solution to disrupt Si-O-Ce bonding/electrostatic interactions and solubilize refractive CeO_2
2) Improve methodology for delivering cleaner
3) Prevent particle redeposition onto the wafer

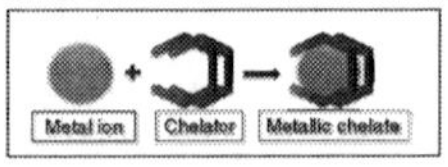

Cleaner development for Ce chelation and dissolution

→ *Ce ion chelation (via components in cleaner) assists in disrupting Si-O-Ce bonding and electrostatic forces, dissolution, and preventing redeposition.*

- Chelation yield determined by titration
- Highest chelation of Ce ions allows for greatest sequestration of Ce residue in aqueous media
- High chelation across a broad pH range enables more versatile applications for cleaner

Complexation with Ce ions across pH range

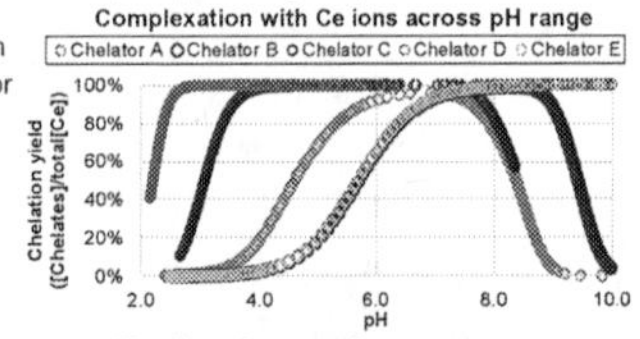

➢ ***Chelator D enables most effective clean performance***

Results and Discussion

Table-top buff cleaning results using DuPont CMP pads and formulated cleaner

- Cleaning efficacy is evaluated by [Ce] remaining on coupon surface after PCMP buff clean compared with that after CMP polish
- [Ce] determined by ICP-MS analysis of digested coupon

Process & Consumables	
Polish tool	Table-top polisher
Buff pad	Commercial pads A, B, C Experimental pad D
Slurry	Negatively-charged commercial ceria slurry
Clean	Buff clean only

Buff clean evaluation of three different DuPont CMP pads with PCMPSolv™ 2110:

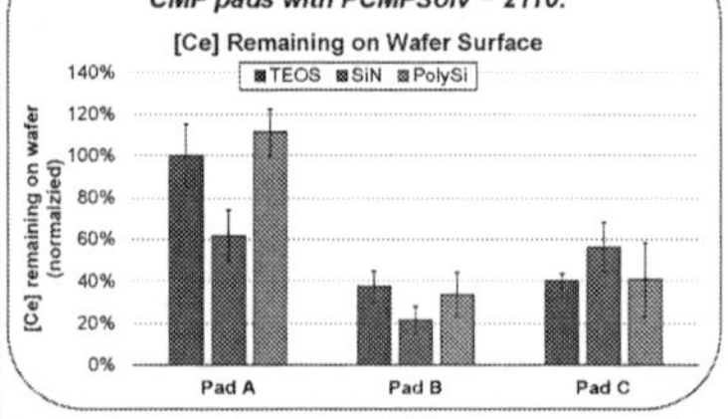

➢ ***On-platen buff clean w/ CMP pads↑contact of pad & cleaner w/ wafer surface for↑clean efficiency.***

Improved cleaning performance with PCMPSolv™ 2110 over commercial citric acid-based cleaner

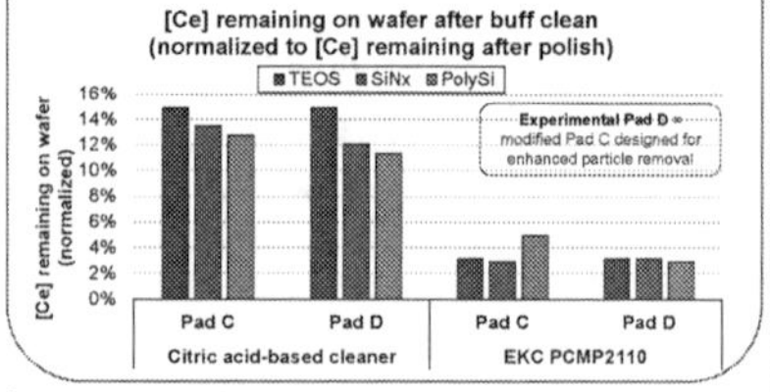

➢ ***Best clean performance is achieved using PCMP2110 with commercial pad C & expt'l pad D.***

Pad evaluation for enhanced ceria removal on 300mm polish tool

Process & Consumables	
Polish tool	300mm polish tool
Polish & buff pad	Pad B Pad C
Slurry	Negatively-charged commercial ceria slurry
Clean (PCMPSolv™ 2110)	Brush clean only or Buff clean + brush clean

- Polish & buff clean steps were done using same pad type
- Pads B and C provided the best cleaning efficacy; however, pad B gave very low TEOS RR

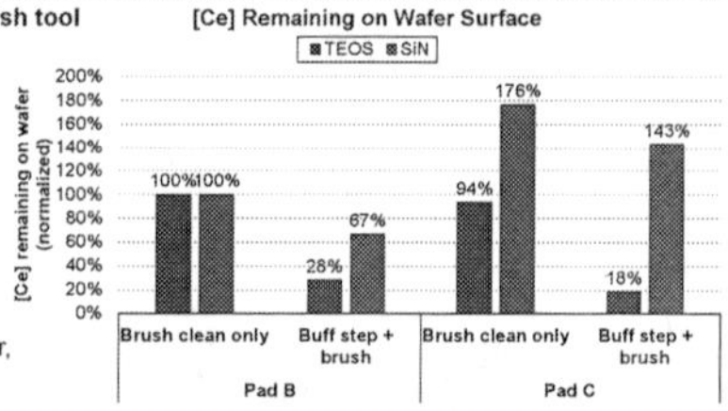

➢ ***PCMPSolv™ 2110 + DuPont pad C combination provides the best overall performance.***
➢ ***Buff clean step significantly improves Ce removal on both TEOS and SiN_x wafers.***

Conclusions and Future Directions

PCMP ceria cleaning is enhanced by delivering formulated cleaner PCMPSolv™ 2110 using DuPont CMP pads, likely owing to the ability of these pads to deliver a higher volume of cleaner to the wafer surface for an extended amount of time. Pad C provides the best combined polishing and cleaning performance. Experimental pad D targets and improves ceria removal from the polysilicon wafer surface. Overall, PCMP2110 facilitates improved cleaning over both commodity-type chemicals and DI water. Future directions will focus on evaluating other DuPont buff pads and formulated cleaners to expand cleaning efficacy. Chemical interactions between ceria particles, pads, and clean components will be further analyzed to both identify existing and develop new cleaners and pads that can leverage these interactions to work synergistically.

June 2021 DuPont Electronics & Industrial

New Formulation for Improved Polysilicon Post-CMP Clean

DUPONT

Ping Tzeng*, Kathryn Gramigna, Wayne Huang, Paul Bernatis, Ling Chang, Adam Tang, Vicky Wu and Kyoung Sik Chin
ACT Technology, DuPont

1 Introduction

After the planarization process, the surface of polysilicon is predominately hydrogen-terminated and hydrophobic. Traditional aqueous cleaners demonstrated inefficient abrasive removal due to their poor wetting capabilities; other challenges such as morphology control (roughness and scratches) and trace metal elimination should be considered as well. All these technical hurdles prompted extensive investigations into formulated cleaner. In this study, we will demonstrate a next-generation solution with improved polysilicon protection and wettability by surface modification that enhance polysilicon CMP cleaning.

2 Fundamental

(i) Excellent wetting capability:

- Enhanced "flowability" to penetrate into the interface between polySi surface and abrasive
- Improved organic residue removal

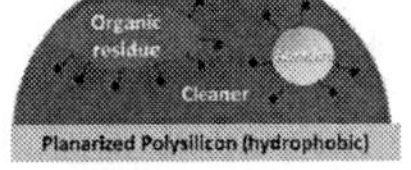

(ii) Surface modification:

- Electrostatic repulsive force to prevent particle re-deposition.
- Treated polysilicon surface becomes hydrophilic, increasing brush clean efficiency.

(ii) Improved post-alkaline compatibility:

- Prevents polySi damage from brush chemical (alkaline).

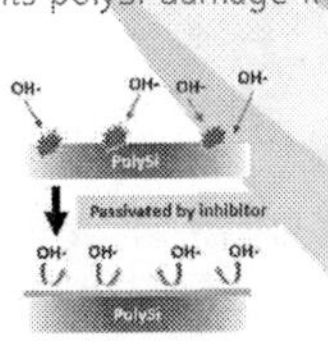

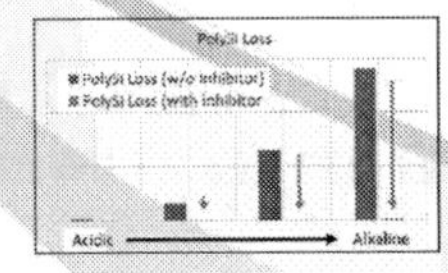

3 Results and Discussion

Formulated Chemistry Makes Synergy Possible

Wettability + Surface Hydrophilicity + Post-alkaline Compatibility = PCMPSolv™ 1100 (neutral)

Contact angle of Cleaner	
EXP-477	PCMPSolv™ 1100
52°	33°

Contact angle of DIW	
EXP-477	PCMPSolv™ 1100
50°	34°

PolySi Loss	
EXP-477	PCMPSolv™ 1100
1900 ppb	340 ppb

PCMPSolv™1100 showed promising cleaning Performance

(a)

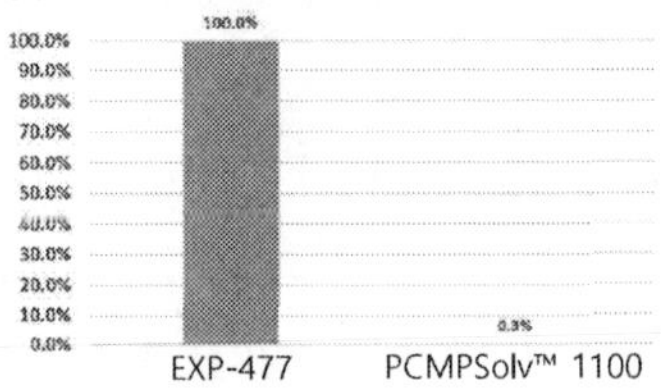

(b)

(c)

Figure 1. (a) Normalized defectivity of different buffing chemical with NH_4OH brush, obtained by KLA SP5; **(b-c)** Haze count obtained by KLA SP5, representing surface roughness.

5 Conclusions or Future Application

Currently, the cleaning process for polysilicon CMP is mainly done by bulk chemicals such as ammonia or SC1. Through our inhouse study, these chemicals can only enable electrostatic repulsion and might cause polysilicon surface damage . Our ACT formulating cleaner applied prior brush clean can easily enhance cleaning efficiency and control surface roughness, which has been demonstrated externally as well.

ACT DuPont RESTRICTED Ping.Tzeng@dupont.com

HUNTSMAN

Enriching lives through innovation

Synthesis and Applications of Quaternary Ammonium Hydroxides for Semiconductor Processes

Hui Zhou[1], Aiyu Yan[2], Victoria White[1], Ke Zhang[1], Jay C. Henderson[1], Graham Gibson[3] and Tim Clark[4]*

[1]: Huntsman Corporation, 8600 Gosling Road, The Woodlands, TX 77381, USA

[2]: Huntsman Shanghai Campus, 479 Wen Jing Road, Minhang Econ. & Tech. Dev. Zone, Shanghai, 200245, CHINA

[3]: NanoFabrication Kingston, Queen's University, 945 Princess Street, Kingston, ON K7L 0E9, CANADA

[4]: GreenCentre Canada, 945 Princess Street, Suite 105, Kingston, ON K7L 0E9, CANADA

*: hui_zhou@huntsman.com

Introduction

Quaternary ammonium compounds (QACs), sometimes referred to as quats, are used in a broad range of industrial applications as disinfectants, surfactants, fabric softeners and antistatic agents. Among them, tetramethylammonium hydroxide (TMAH), is commonly used in semiconductor applications such as photoresist developing, post etch residue removal or post chemical mechanical planarization cleans. However, TMAH is known to have toxicity concerns, even in very diluted solutions, resulting in an increasing demand for TMAH alternatives with improved safety and hazardous waste handling.

Huntsman has developed a family of quaternary ammonium hydroxides, such as E-GRADE® Choline OH, E-GRADE® THEMAH, XHE-124, XHE-125, and XHE-128 that offer comparable performance and safer handling while providing our customers alternatives for their formulations. As an amine supplier we offer the ability to source quats from a broad range of amine compounds offering unique structures and properties. Our quaternary amines feature -CH_2CH_2-OH functionalities offering unique solubility and surface interactions and are available in higher concentrations than TMAH.

Applications: Surface texturization of Si wafer for Photovoltaic Applications

- Texturization reduces light reflection and improves light absorption (Figure 1)
- Etching with quats can generate different surface textures
- Both E-GRADE® Choline OH and E-GRADE® THEMAH showed better uniformity (Figure 2)
- Optimized by temperature, time, and concentration

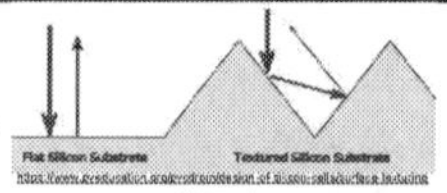

Figure 1

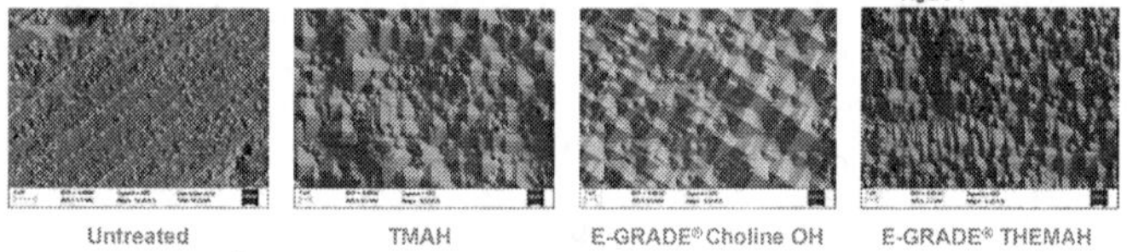

Figure 2. SEM photos of Si wafer surface

- Texturized Si wafer (treated with E-GRADE® Choline OH) has significant reflectivity reduction compare to untreated substrate (Figure 3)
- 20-30% reflectivity reduction has been observed
- Si wafer etched by E-GRADE® Choline OH showed lower reflectivity and more homogeneity than those treated with TMAH (Figures 2 and 4)

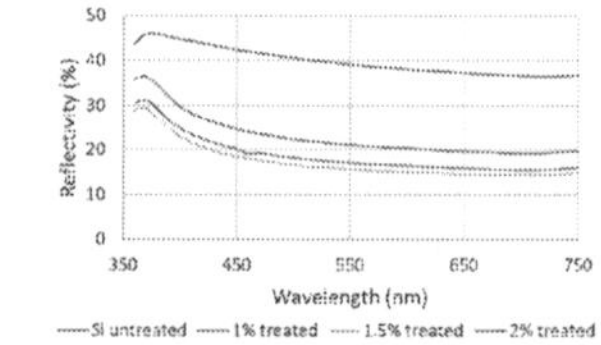

Figure 3

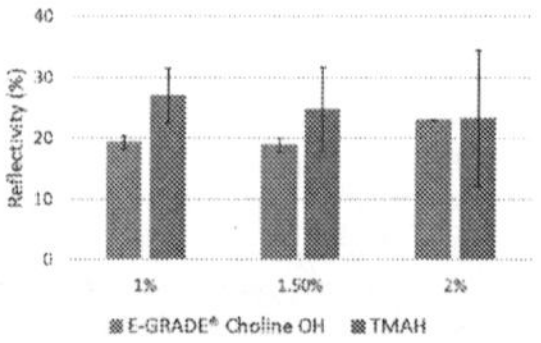

Figure 4

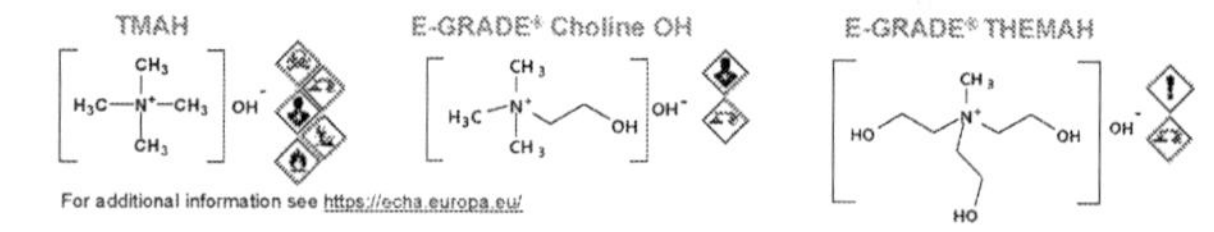

For additional information see https://echa.europa.eu/

- E-GRADE® Choline OH and E-GRADE® THEMAH are strong quaternary ammonium salts with lower toxicity

Applications: Photoresist development

Photoresist Removal

- Quats are a key component in photoresist developer formulations
- Si wafers were spin-coated with commercially available positive photoresist, exposed to patterned UV light, and developed in three developer solutions
- TMAH showed faster removal rate than E-GRADE® Choline OH and E-GRADE® THEMAH at same molar concentration (Figure 5)
- TMAH, E-GRADE® Choline OH and E-GRADE® THEMAH have good edge development (Figure 6)

Unexposed Photoresist Removal

- In another study, unexposed positive photoresist coated Si wafers were soaked in solutions of each of the three quats
- E-GRADE® Choline OH exhibited faster removal rate than TMAH at 20% solution, which can be an option used in stripping process

Metal corrosion

- Etching rates were measured by Linear Polarization Resistance technique (LPR) at room temperature
- E-GRADE® Choline OH and E-GRADE® THEMAH exhibited lower or comparable corrosion rate than TMAH in most cases
- The combination of low corrosion rate, good photoresist removal, as well as lower toxicity makes E-GRADE® Choline OH and E-GRADE® THEMAH good alternatives to TMAH

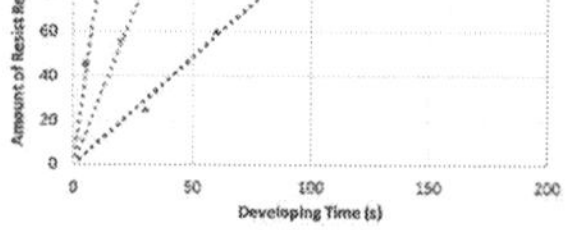

Figure 5

Figure 6. SEM photos after 100% removal

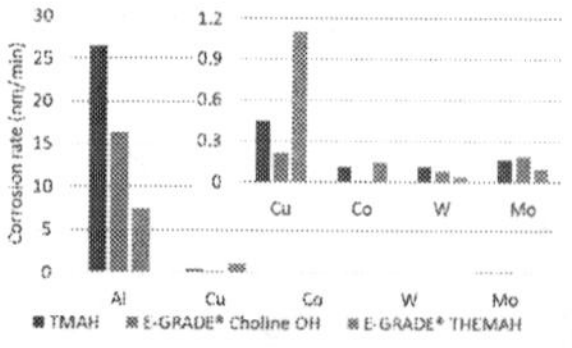

Figure 7

Summary

A family of quaternary ammonium hydroxides, such as E-GRADE® Choline OH, E-GRADE® THEMAH, XHE-124, XHE-125, and XHE-128, offers a variety of structures, tunable performances, and lower toxicity options for the semiconductor industry.

SPCC, June 8-10, 2021, USA

AUTHOR INDEX